Where Does the W

DATE DUE

Where Does the Weirdness Go?

WHY QUANTUM MECHANICS IS STRANGE, BUT NOT AS STRANGE AS YOU THINK

DAVID LINDLEY

BasicBooks
A Division of HarperCollinsPublishers

530.12
L746w

Designed by Elliott Beard

Library of Congress Cataloging-in-Publication Data
Lindley, David, 1956–
    Where does the weirdness go? : why quantum mechanics is strange, but not as strange as you think / by David Lindley.
        p.   cm.
    Includes index.
    ISBN 0-465-06785-9 (cloth)
    ISBN 0-465-06786-7 (paper)
    1. Quantum theory.   2. Physics—Philosophy.   I. Title.
QC174.12.L54   1996
530.1'2—dc20   96-1049
                                                      CIP

97 98 99 00 ❖/RRD 9 8 7 6 5 4 3 2 1

Contents

Intermission: A Largely Philosophical Interlude

Act II: Putting Reality to the Test

Act III: Making Measurements

Introduction: Why do I trust my computer?

The computer I've been using to write these words has been satisfactorily reliable: I switch it on and off repeatedly, calling up files that contain the words I wrote last time, adding new words, shuffling the old ones around, and saving the results for next time. I rarely trouble to think what is going on inside the computer that lets me see my words on the screen, or move them painlessly from one place to another, or restore a sentence that I accidentally erased, or play a game of solitaire in the odd moment when inspiration deserts me. And if I do think about these inner workings, I'm not nearly enough of a computer expert to be able to say at all accurately what is happening in the machine. Instead, I tend to comfort myself with plausible analogies that give me a sense that I more or less know how the computer works, without going to the difficulty of mastering the volumes of technical detail I would need to know to understand it properly (which, I'm happy to say, I don't need to. The reliability of my computer gives me ample confidence that there are dogged and knowledgeable people in the world who can indeed design and build these things).

At the bottom of it all are electric currents carried by microscopic charged particles called electrons. Rattling around in my computer, I like to think, are little streams and packages of electrons that constitute the electrical signals, the binary zeroes and ones that form the basis of its inner workings. Somehow, the letters on the screen are built from patterns of electrical signals, and somehow, my instructions to the computer from the keyboard cause these patterns of electrical signals to change and move. So I think of the computer as a vast, intricate electronic pinball machine, with unimaginable numbers of pathways and trajectories, and with exquisitely timed and delicately adjusted flippers that guide electrons this way and that to produce a constantly changing, frenetically busy but nevertheless consistently and accurately meaningful pattern of electronic flows. The reliability and precision of all this activity, despite its daunting complexity, is the truly stunning part of computer design, and that's the bit I don't pretend to understand. My words take shape as a buzzing pattern of circulating electrons, and that's about as much as I want to know.

And when I have done for the day and want to store what I have written, I can tell the computer to send the sequence of electrical zeroes and ones to the hard disk, where they are encoded now as a series of magnetic blips on the disk's surface. To get an idea of how the hard disk works, I imagine its surface to be studded with tiny magnets whose poles can be flipped one way or the other on command, to record either a zero or a one. The hard disk is perhaps ten centimeters across, and can store 120 megabytes of data (the computer is a few years old, or that figure would be more like 1,000 megabytes); one byte, in standard computer technology, is a word of eight binary bits—eight zeroes or ones—so that all in all my hard disk can accommodate close to a billion blips of data. Each of those tiny magnets must, according to a quick calculation, be a few millionths of a meter across. This is the size of a grain of dust, too small to be seen by the unaided eye, and yet my computer can record and retrieve data on the hard disk as if these magnetized dust grains were levers that could be set firmly up or down, like the signal levers that an old-time railway signalman would operate, and it can set and read millions of these levers in a fraction of a second. How can invisible dust grains be so dependable? How can I store and retrieve a file of written words hundreds of times without ever a single dust grain accidentally flipping the wrong way, or being disturbed by some wayward external influence?

On the rare occasions that I think about the inner workings of my computer, I resort to mechanical images of this sort. I conjure up familiar pieces of machinery—pinball flippers, railway switches and signals—and then imagine that these devices can be reduced to the size of dust grains, and arranged into fantastically complicated networks. This doesn't really tell me how my computer works, but it lets me think I have the right kind of idea in my head, and that I could understand it, really, if I wanted to.

But then, in another part of my mind is the recollection of undergraduate physics lectures in which I learned that electrons are fundamentally not at all like pinballs. There was something called the uncertainty principle of quantum mechanics, which says that you can never know exactly where a microscopic particle such as an electron is at any one time, or how fast it is moving, so that if you want a picture of an electron you have to think, perhaps, of a blurry, out-of-focus, smudged-out pinball. And there was another puzzling idea, by the name of wave-particle duality, according to which an electron can behave sometimes in ways that make you think it is a particle, but at other times in ways that make you think more of waves. It is both wave and particle, or perhaps neither wave nor particle but something in between, undefinable and unimaginable; at any rate, even the idea of a smudged-out pinball begins to seem dubious. And on top of all that there was a vague notion about measurements affecting in unpredictable ways the things you are trying to measure, so that even if you have a device that can tell where one of these smudged-out, wavey-particley things is, you can't quite be sure of the meaning or reliability of the answer you get.

And now, thinking about all this, my assurance that I understood how my computer works and how it can be so reliable begins to crumble. If I'm not allowed to think of the electrons as pinballs rattling around the precisely engineered pathways of the silicon chips, if they are really sloshing about like waves in channels, if the uncertainty principle tells me an electron cannot be altogether in this place but has to be also a little bit in that place at the same time, how can my computer perform the same tasks over and over again with such reliability? And if there's some unpredictability associated with every act of measurement, how can I trust the data I read off the hard disk since, in effect, reading the data amounts to measuring the orientations of all those little magnetized dust grains? Quantum mechanics, or so I recall from my education in physics, says that at the most fundamental level, the world is not wholly knowable, and not wholly dependable. In dealing with individual electrons or the magnetic alignment of individual atoms, I must think not in certainties but in probabilities.

Nevertheless, my computer continues to work, as imperturbably as ever. A standard answer to this riddle is that, in fact, a computer does not rely on individual electrons and atoms for its operation. The signals that make up the zeroes and ones chasing around its silicon pathways are gangs of perhaps a trillion electrons. The magnetic dust grains on the hard disk are built from trillions of atoms. These things may be microscopic by human standards, but compared to the individual inhabitants of the quantum world they are nevertheless gigantic. And so, it's sometimes claimed, the quantum mechanical strangeness that besets individual electrons and atoms, and bedevils our thinking about them, becomes negligible when we think about the trillions of electrons and atoms on whose collective behavior my computer depends.

But what sort of an answer is this? Why should an assembly of a trillion weird little quantum objects behave any less mysteriously than its components? A trillion drops of water make a bucket of water, not a concrete block. If it's true that the weirdness of the quantum mechanical world seems to disappear when we look at "big" objects, then where, precisely, does that weirdness go? If we can't trust a single electron to be precisely in one place at one time, how can we trust a throng of electrons to invariably represent the letter a on my computer screen, and not turn casually into a z?

For many decades, this question was resolved by flat assertion. It was simply declared that any measurement produced, of necessity, a definite answer, and thereby forced definition onto the uncertain, ambiguous quantum world. But what a measurement was, by what physical process it made indefinite things definite, was never accounted for. In the last few years, however, the beginnings of an answer to this long-standing puzzle have begun to appear. The answer derives, in part, from theoretical insights into the behavior of complex systems, which have made it possible to understand how assemblies of many interconnected quantum objects can behave in collective ways that are by no means obvious, or easily deduced, from the behavior of those single objects in isolation.

The purpose of this book is to explain this new understanding. We will see that although the weirdness does not altogether go away, it does fade into the background.

To understand the answer, you have to first formulate the question. The quantum world is an undeniably strange place, working to unfamiliar rules, and in the first part of the book I have tried to explain, as clearly as I know how, what that strangeness consists of and (just as important) what it is not. With the essentials laid out, I delve briefly into some of the misguided efforts that have been pursued over the years in the hope of making quantum mechanics look less weird than it really is. Only, in the end, by accepting the true nature of quantum mechanical weirdness does it become possible to see exactly what the central problem is, and how, in practice, nature gets around it.

The book is organized in what I hope is a logical rather than a chronological manner. I have plunged in at the beginning with one of the well-established and much-discussed "paradoxes" of quantum mechanics, and tried to work from there to an understanding of why the paradox arises. The book's organization seems logical to me, anyway. The reason quantum mechanics is disconcerting is that it seems to make nonsense of our usual definitions of logic, leaving us with nothing to hang on to. But read on: in the end, logic reappears, and the world makes sense again!

Acknowledgments

For comments on the manuscript I am grateful to Bob Shackleton and to Susan Rabiner, my editor at Basic Books. For the illustrations, thanks to Liz Carroll and Preston Morrighan.

For companionship, thanks finally to B.C., who did her best not to walk across the keyboard too often, and who remained until the end happily unaware of Schrödinger's conundrum. Alter ipse amicus.

Dramatis Personae

Niels Bohr—a sage, late of Copenhagen; the founding father and guiding spirit of the Copenhagen interpretation of quantum mechanics

Albert Einstein—physicist, father of relativity, godfather to quantum mechanics, though later estranged therefrom

Erwin Schrödinger—owner of a cat, though not necessarily a cat-lover

Max Planck—originator, arguably, of quantum mechanics, though he sought in vain to disown his offspring

David Bohm—heir to Einstein's mantle, who sought to install quantum mechanics on a classical foundation, and not vice versa

John Bell—a sympathizer of Einstein and Bohm, who devised a test the outcome of which would have disheartened Einstein

Copenhagen—city of Denmark; also, a stern philosophy

Electron—an elementary particle, of fixed mass and electric charge, discovered in 1897; later found also to be a wave

Photon—a particle; also, a wave

ACT I

Mechanical Failure

From the days of Newton and Descartes up until the end of the nineteenth century, physicists had constructed an increasingly elaborate but basically mechanical view of the world. The entire universe was supposed to be a glorious clockwork, whose intricate workings scientists could hope to find out in limitless detail. By means of the laws of mechanics and gravity, of heat and light and magnetism, of gases and fluids and solids, every aspect of the material world could in principle be revealed as part of a vast, interconnected, strictly logical mechanism. Every physical cause generated some predictable effect; every observed effect could be traced to some unique and precise cause. The physicist's task was to trace out these links of cause-and-effect in perfect detail, thereby rendering the past understandable and the future predictable. The accumulation of experimental and theoretical knowledge was taken unarguably to bring a single and coherent view of the universe into ever sharper focus. Every new piece of information, every new intellectual insight, every new elucidation of the linkages of cause-and-effect added another cog to the clockwork of the universe.

This was the tradition in which physicists at the end of the nineteenth century had been raised. Classical physics aspired to portray with perfect clarity the intricate workings of the mechanical universe. That the real universe was indeed mechanical, that physicists were depicting in ever sharper focus a reality that existed independently of them—these self-evident suppositions were never questioned.

The mystery of the other glove

You and a friend are at Heathrow Airport, London. You each have a locked wooden box containing a glove. One box contains the right-handed glove of the pair, the other the left-handed glove, but you don't know which box is which. Both of you also have keys, but they are not the keys to the boxes you are carrying.

Thus equipped, you board a plane and fly to Los Angeles. Your friend flies at the same time to Hong Kong.

When you get to Los Angeles you use your key to open a locker at the airport, and inside you find another key. This is the key to your wooden box, which you now open to discover that the glove you have brought to Los Angeles is the right-handed one. As soon as you know this, of course, you know also that your friend's wooden box, by now in Hong Kong, contains the left-handed glove. With that instantaneous realization, you have acquired a piece of knowledge about a state of affairs on the other side of the world.

Perfectly straightforward, you may say, and so it is. You may have heard of Albert Einstein's famous dictum that nothing, not even information, can travel faster than the speed of light, but no part of this little screenplay contradicts that injunction. You have indeed made a deduction, using information available to you as you wait at the Los Angeles airport, about a fact that pertains to your friend in Hong Kong. But we make these kinds of long-distance inferences, in big ways and small, all the time. An astronomer catching the feeble rays of light that reach a telescope here on Earth thereby deduces the surface temperature of a distant star. You get out of the shower one morning, look at your watch, and realize that a meeting in your office that you had to attend has already started.

Figuring out what is happening in some distant place is a different thing from transferring that knowledge from one place to another. If, having discovered that your glove is right-handed, you wanted to tell your friend that she has the left-handed one, you would have to pick up the phone, or send a telegram, or mail her a postcard. A phone call might travel almost at the speed of light, the other two messages much slower. And you have no way of knowing whether she has already opened her box or not—unless you happen to get a phone call from her telling you that you must have the right-handed glove. The fact that you have found out which glove she has does not allow you to beat the laws of physics and get that information to her faster than Einstein allows.

But still, you think there might be some way of exploiting your knowledge to influence your friend's behavior. Suppose, before you both set off on your plane trips, you had agreed with your friend that if she found the left-handed glove in her box she would proceed to Tokyo, but if she got the right-handed one she would fly to Sydney. Does your opening the box in Los Angeles determine where she ends up? By no means: whichever glove was in her box was there from the outset, so whether she has to fly to Tokyo or Sydney is predetermined. When you open your box in Los Angeles you instantly know where she must be going next, but her destination is as much of a surprise to her as it is to you. As before, you've now found out what happens next, but you haven't had any influence over it.

But now let's change the story. The gloves in the two boxes are, you are informed, of a strange and magical kind, unlike any gloves you have come across before. They still make up a pair, but for as long as they are sealed in their boxes, they are neither right-handed nor left-handed but of an unfixed, indeterminate nature. Only when a box is opened, letting in the light, is the glove inside forced to become either right-handed or left-handed. And there is a fifty-fifty chance of either eventuality.

During the several hours you are in the plane flying from London to Los Angeles, you may well be puzzling over what the glove in your box—this strange glove, neither right-handed nor left-handed but potentially either—actually looks like. But you don't have the key that would let you open the box and peek inside, and in any case, as soon as you peeked the glove would have to take on a definite shape, right-handed or left-handed. The magical nature of this glove is such that you can never see it in its unformed state, because as soon as you look, it turns into something familiar and recognizable. A frustrating catch-22.

On the other hand, when you now arrive at Los Angeles and open your box to find, let us suppose, a right-handed glove, you begin to think that things are not as straightforward as before. You immediately know that when your friend opens her box, she must discover a left-handed glove. But now, apparently, some sort of signal or information must have traveled from your glove to hers, must it not? If both gloves were genuinely indeterminate before you opened your box and looked inside, then presumably as soon as your glove decided to be a right-handed one, hers must have become left-handed, so that the two would be guaranteed to remain a pair. Does this mean that your act of observing the glove in Los Angeles instantaneously reduced the indefiniteness of its partner in Hong Kong to a definite state of left-handedness?

But it occurs to you that there's another possibility. How do you know your friend didn't get to Hong Kong first and open her box before you had a chance to open yours? In that case, she evidently found a left-handed glove, which forced yours to be right-handed even before you looked inside your box. So if there was an instantaneous transmission of information, it might have gone the other way. Your friend's act of opening her box-determined what sort of glove you would find, and not the other way around.

And then, you think, the only way to find out which way the instantaneous information went, from your glove to hers or from hers to yours, is to pick up the phone, call Hong Kong, and find out what time she opened her box. But that phone call goes no faster than the speed of light. Now you are getting really confused: there seems to have been some kind of instantaneous communication between the two gloves, but you can't tell which way it went, and to find out you have to resort to old-fashioned, slower-than-light means of communication, which seems to spoil any of the interesting tricks you might be able to figure out if there really had been an instantaneous glove-to-glove signal.

And if you think again of the strategy whereby your friend had to get on a plane to either Tokyo or Sydney, depending on which glove she found in her box, you realize you are no more able than before to influence her choice by your action in Los Angeles. The rules of the game are such that you have a fifty-fifty chance of finding either a right-handed or a left-handed glove in your box, so even if you are sure that you have opened your box before she opened hers, and even if you think that opening your box sends an instantaneous signal to hers, forcing her glove to be the partner of yours, you still have no control over which glove you find. It remains a fifty-fifty chance whether she'll end up in Tokyo or Sydney, and you still have no say in the matter.

And now you're even more confused. You think there's been some sort of instantaneous transmission of information, but you can't tell which way it went, and you can't seem to find a way to communicate anything to your friend by means of this secret link between the gloves.

And perhaps you conclude it's a good thing real gloves aren't like this.

•

In that, you would be in agreement with Albert Einstein. It's true that gloves don't behave this way but, according to quantum mechanics, electrons and other elementary particles do. These particles have properties which, apparently, lie in some unresolved intermediate state until a physicist comes along and does an experiment that forces them to be one thing or the other. And that physicist cannot know in advance, for sure, what particular result any measurement is going to yield; quantum mechanics predicts only the probabilities of possible results.

This offended Einstein's view of what physics should be like. Before quantum mechanics came along, at the beginning of this century, it was taken for granted that when physicists measure something, they are gaining knowledge of a preexisting state. That is, gloves are always either right-handed or left-handed, whether anyone is looking at them or not, and when you discover what sort of glove you have, you are simply taking note of an independent fact about the world. But quantum mechanics says otherwise: some things are not determined except when they are measured, and it's only by being measured that they take on specific values. In quantum mechanics, gloves are neither right-handed nor left-handed until someone takes a look to find out. At least, that is what quantum mechanics seems to say.

The story we just went through, about indeterminate gloves being taken to separate places and examined by two different people, is part of an experimental setup that Einstein and some colleagues devised as a way to show how absurd and unreasonable quantum mechanics really is. They hoped to convince their physicist colleagues that something must be wrong with a theory that demanded signals traveling faster than the speed of light.

But, as the Danish physicist Niels Bohr was quick to point out, it's far from clear if anything genuinely unacceptable has actually happened with these magical gloves. The whole thing may seem very odd, and it may seem quite inescapable that some sort of instantaneous communication between the gloves is essential for the trick to work, but in the end it seems impossible to do anything with the supposed communication. Bohr arrived at what he deemed an acceptable interpretation of this sort of puzzle by forcefully insisting that one must stick to practicalities: it's no good, and indeed positively dangerous, to speculate about what *seems* to happen in such a case; stick to what actually occurs, and can be recorded and verified, and you'll be all right. If you can't actually send an instantaneous message of your own devising, then it's meaningless to guess at what might or might not have been furtively going on between the two gloves.

Nevertheless, Einstein persisted in objecting to what he called this "spooky action-at-a-distance": action-at-a-distance because an occurrence in one place seems to have an instantaneous effect somewhere else, but spooky because the influence is implied rather than directly seen. Einstein accepted, more or less, Bohr's argument that you could stay out of trouble by sticking with documented and unambiguous facts, but to him this was a philosophy that worked only if you were willing to deliberately blind yourself to deeper issues. And many physicists and philosophers since then have found themselves dissatisfied by Bohr's workable but minimalist views.

To understand these disputes we need to take our gloves off and come to grips with the essentials of quantum mechanics. A good place to start is this matter of things being indeterminate until measured. What does this mean, and where does it come from?

In which things are exactly what they are seen to be

Ultimately, there must be recourse to experimental evidence. If quantum mechanics asserts that the act of measurement does not simply yield information about a preexisting state, but rather forces a previously indeterminate system to take on a definite appearance, there must be empirical reasons for the assertion. Even theoretical physicists would not come up with so bizarre and counterintuitive an idea if they were not forced to it.

One of the first such demonstrations, and still perhaps the easiest to grasp, was performed in Germany, in 1921, by Otto Stern and Walter Gerlach. For the purposes of this experiment, think of atoms as little bar magnets, with north and south poles (where the magnetism comes from has to do with the nucleus of the atom and the electrons orbiting it, but we don't need to worry about that here).

What Stern and Gerlach did, in essence, was to send a beam of atoms through a region of magnetic field, and record in what direction the atoms came out. Now, if little bar magnets are sent through a completely uniform magnetic field (whose intensity, in other words, is the same everywhere) nothing interesting will happen. Whatever upward force might be exerted on one pole of the magnet will be exactly balanced by a downward force on the other pole, and the magnet as a whole will sail through undisturbed.

So Stern and Gerlach arranged that their magnetic field should be graded in intensity, let us say from top to bottom. In this case, a bar magnet that is aligned vertically will feel a stronger force on its upper end than its lower, and will move either up or down depending on which of its poles is at the top (for example, if the magnet has its north pole upward, and the magnetic field it's passing through has a south pole towards the top, and is strongest there, it will pull the north pole upwards more strongly than it pushes the south pole downwards, and the whole thing will drift upwards). A bar magnet that travels through the magnetic field horizontally will travel undisturbed, because both its poles will experience the same intensity of field, and up and down forces will cancel.

These details aren't important. The only point to grasp is

that Stern and Gerlach used a magnetic field whose graded intensity was designed to sort atoms according to their magnetic orientations. Atoms pointed one way move upwards, atoms pointed the opposite way move downwards, and atoms in intermediate orientations are deflected by the appropriate intermediate amount. The experiment then consists of shooting atoms through the magnetic field and recording their positions as they come out the other end—for example, by letting them hit a photographic film, or a phosphorescent screen that records a bright spot every time an atom strikes it.

What do we expect to see? When we shoot a beam of atoms through the device, we expect their magnetic axes to point at random, in all possible directions. If that's the case, then a few atoms should suffer the maximum deflection upwards, a few the maximum deflection downwards, while others should go through at all intermediate deflections. The atoms in the beam should therefore be smeared out, when they hit the screen, into a range of all possible deflections.

This is a straightforward and seemingly unarguable result, but it is not what Stern and Gerlach found. Instead, they discovered that the atoms emerging from their magnet struck the screen in only two places, having been deflected by equal amounts either up or down. It was as if the little atomic magnets, rather than adopting any random orientation, were obliged to line up either parallel or antiparallel to the magnetic field—that is, either vertically up or vertically down. (See Figure 1.)

Moreover, if the Stern-Gerlach magnet itself was flipped through ninety degrees, so that its magnetic field was horizontal instead of vertical, the atoms struck the screen in two spots to the left and right of the original line of the beam.

In fact, Stern and Gerlach discovered, no matter how the magnetic field was lined up, it always split the beam of atoms into two, each atom being forced somehow to take up either one or the other of just two possible orientations. And these possible orientations were dictated, apparently, by the align-

FIGURE 1

What you'd expect . . .

What you get . . .

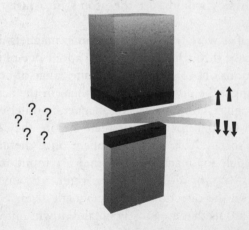

In a classical experiment, little bar magnets would be deflected by the magnetic field of a Stern-Gerlach device according to their fixed but unknown orientations. When the experiment is done with atoms, however, the device sorts them into only two paths, with fifty-fifty probability.

ment of the Stern-Gerlach magnet, not by anything to do with the atoms themselves.

Now, you might imagine that somehow the interaction between the little atomic bar magnets and the field of the Stern-Gerlach magnet is more complicated than we have so far supposed. It might be, for example, that somehow, as they pass through the magnetic field, the atoms find that only two orientations out of all the possible orientations are stable, so that no matter what direction the atoms' magnetic axes were pointing in before they entered the magnetic field, they are twisted and rotated around inside the device so that they emerge with only two possible orientations.

There are two problems with this. First, on theoretical grounds, what happens between the atoms and the Stern-Gerlach magnet is basically a matter of elementary electromagnetic theory and mechanics, and there is no phenomenon by which classical physics would sort the atoms into just two stable orientations.

Second, if it were true that the atomic magnets are forced, during passage through the magnet, to adopt one of two possible orientations out of an initially infinite range of possibilities, then it ought to be possible to catch them in the act. Pass the atoms through a shorter or weaker magnet, in other words, and you should be able to intercept them before they have been completely sorted. This doesn't happen: once the atoms have passed through any magnetic field, they are found in only two orientations—up or down, left or right, or anything else depending on the direction of the magnetic field. No one has ever found atoms that are mostly up and down, but with a few in between that haven't been properly sorted yet.

This result, simple and robust, has become a secure part of the empirical foundations of quantum theory. How are we to interpret it? We still want to believe that the atoms going into a Stern-Gerlach magnet must have all possible orientations, because no matter how the magnet is set up (vertical, horizontal, or anything in between), the result is the same: the beam gets split into two equal parts. Therefore, nothing about the original atomic beam can have any preferred orientation in one direction or another.

On the other hand, we have also satisfied ourselves that the Stern-Gerlach magnet does not physically "sort" the atoms passing through by directly manipulating their magnetic axes. These two statements seem to be contradictory. If all possible atomic orientations are present in the incoming beam, and if the magnet itself does not physically shift those orientations, then we ought to get all possible atomic deflections at the other end of the magnet. And yet, in every case, only two possible orientations emerge.

In classical physics this is indeed an unresolvable contradiction. That's one of the reasons quantum mechanics is fundamentally different. In quantum theory, the Stern-Gerlach experiment is explained like this: passing an atom through a magnetic field amounts to a measurement of its magnetic alignment, and until you make such a measurement you have no business saying what the atom's magnetic alignment might be; *when* you make a measurement, however, you obtain one of only two possible outcomes, with equal probability, and those two possibilities are defined by the direction of the magnetic field that you use to make the measurement.

Did we leave something out? What happened to the statement that before measurement the atoms had random orientations? But this is precisely the point at which we have to change our thinking. In classical terms, "random orientations" is taken to mean that each atom has a definite but unknown magnetic alignment, and that taken as a whole, the beam includes all possible alignments. But this is what got us into trouble, because it would mean that the outcome of a Stern-Gerlach measurement would likewise have to include all possible values, not just two. Instead, we have to say that the magnetic orientations of the atoms are indeterminate, rather than random, before any measurement is made. "Indeterminate" means just what it says—that the atoms cannot be assigned any meaningful magnetic orientation, not that each of them has some orientation that we don't happen to know.

For some purists, in fact, even saying that the magnetic orientations are indeterminate is a little too assertive; the puritanical thing to say is that until you make a measurement of magnetic orientation—for example by, passing an atom through a Stern-Gerlach magnet—the very term has no meaning. "Magnetic orientation" should be construed, in this new language, not to mean "a certain atomic property, indeterminate or otherwise, that can be found out by a suitable measurement" but rather "the result that's obtained when a measurement of magnetic orientation is made." This sounds circular: what it really means is that in quantum mechanics, a measurement means precisely and only the result of an act of measurement.

The sparest summary of what happens in a Stern-Gerlach magnet is that when a measurement of magnetic orientation is performed, two outcomes are possible. Magnetic orientation is not defined, in a pragmatic sense, except by making a measurement of it, so that even to say the orientation is indeterminate or indefinite before the measurement is to attribute some sort of reality or credibility to a notion—magnetic orientation—which strictly speaking is meaningless except as it is defined by the process of measuring it.

This is the heart of a fundamental issue. In classical physics, we are accustomed to thinking of physical properties as having definite values, which we can try to apprehend by measurement. But in quantum physics, it is only the process of measurement that yields any definite number for a physical quantity, and the nature of quantum measurements is such that it is no longer possible to think of the underlying physical property (magnetic orientation of atoms, for example) as having any definite or reliable reality before the measurement takes place.

To put it another way: in classical physics, we conventionally think of a physical system as having certain properties, and we imagine—or actually perform—experiments that provide us with information about this preexisting system. But in quantum physics, it is only the conjunction of a system with a measuring device that yields definite results, and because different measurements (applying a Stern-Gerlach magnet with either up-down or left-right orientation, for example) produce results that, taken together, are incompatible with the preexistence of some definite state, we cannot usefully define any sort of physical reality unless we describe not only the physical system under scrutiny but also, and with equal importance, the measurements we are making of it.

This, no doubt, is baffling. We are, through long familiarity, grounded in the assumption of an external, objective, and definite reality, regardless of how much or how little we actually know about it. It is hard to find the language or the concepts to deal with a "reality" that only becomes real when it is measured. There is no easy way to grasp this change of perspective, but persistence and patience allow a certain new familiarity to supplant the old. Let us explore further.

Block that metaphor!

The Stern-Gerlach experiment was first done using a beam of atoms, which were separated into "up" and "down" streams according to their magnetic orientation. But it was soon learned that all kinds of objects, not just atoms but individual subatomic particles such as electrons, suffer a similar fate when passed through a Stern-Gerlach device. The principle turns out to be a universal one: any beam of objects that could, in the perspective of classical physics, be pictured as little bar magnets is found to separate into distinct beams rather than being spread out into a continuous range of diverging paths.

You can imagine the magnetic character of an atom being generated by the swirling motion of electrons orbiting around the nucleus, just as electrons constituting an electric current in a loop of copper wire will produce a magnetic field. But for electrons themselves—tiny individual particles carrying electric charge but with, so far as anyone knows, no internal structure—that picture doesn't seem to apply. Nevertheless, a beam of electrons, just like a beam of atoms, will be split into two by a Stern-Gerlach magnet, separating into "up" and "down" beams.

(At this point we are going to skate over a small technical complication. Because electrons, unlike whole atoms, carry a net electrical charge, any magnetic field will exert a force on them, which is how an electron beam is steered inside your computer monitor or television set to paint a picture on the screen. This force is distinct from, however, and actually at right angles to, the force created by the special magnetic arrangement in a Stern-Gerlach device that acts on the intrinsic magnetic, rather than electric, properties of the electrons, separating them into two beams. In everything that follows we will take it for granted that a suitable device can indeed separate electrons in the desired way, and ignore other electromagnetic effects.)

Loosely, you might think of the electron as spinning like a top, so that somehow its rotating electric charge generates magnetism, and for that reason the quality of an electron that's measured by a Stern-Gerlach magnet—the thing that comes out up or down—is known to physicists as "spin." But picturing "spin" by pretending that the electron is a tiny spinning top gets us into trouble because this quantum kind of spin does not behave, when an electron travels through a Stern-Gerlach magnet, as we would expect a genuine mechanical, classical spin to behave.

And when we try to think of electron "spin" as being in an

uncertain, undefined state—partly up, partly down, as if the little spinning top were somehow capable of being both ways up at the same time—we arrive at a picture that we simply can't imagine. And for good reason: a genuine spin just can't be like this.

If we stick to the barest possible description of the Stern-Gerlach magnet, we need only say that electrons have a certain property that causes them to go one way or the other when traveling through the magnetic field, so that a single beam of electrons is split into two. And we can call the result of this measurement "up" or "down," or "left" or "right," in a perfectly literal way, without ever specifically needing to know what the electron is doing inside the magnet. And since we are not bothering at this point to describe in any more fundamental a way what the electron property is that we are actually measuring (since, to adopt this strict philosophy, it is only as the outcome of some sort of measurement that we can understand what any physical property "is"), it is perhaps a little less baffling to think of the unmeasured electron as having an indefinite value of whatever property it is that we have deliberately failed to define. If you see what I mean.

On the other hand, we need a shorthand way to refer to Stern-Gerlach magnets and the thing they measure, so we will often refer to the spin of an electron as if it were some sort of familiar property. But bear in mind that this kind of spin is not to be taken too literally.

Learning through repetition

One way to make a definitive demonstration of the differences between classical and quantum physics is to take one of the beams of electrons that has emerged from a Stern-Gerlach magnet, and pass it through another such device. Think of a vertically aligned

Stern-Gerlach magnet, and label the two beams that emerge the "up" and "down" beams. Let us from now on think exclusively of electrons in our Stern-Gerlach magnet; because they are truly elementary particles there is no danger of thinking that any of the strange phenomena we might encounter could be due to complicated internal goings-on, which might conceivably be the case with atoms.

We would like to think that the up and down beams each consist of electrons that are all aligned in the same direction, having been forced by the Stern-Gerlach magnet to choose either an up or a down orientation. Is this in fact true?

In one sense, yes. If we take the up beam from a Stern-Gerlach magnet and pass it through another magnet, also vertically aligned, it will receive another little deflection in the up direction, but it will not be split into two. Evidently, these are all "up" electrons, as we had guessed.

But in another sense, they are not. If we take this same up beam and pass it now through a Stern-Gerlach magnet that is set up horizontally rather than vertically, the electrons must be deflected to either the left or the right. We might have liked to think, going back to our classical reasoning, that a beam of up electrons would pass through a horizontal magnet unperturbed, on the grounds that little bar magnets at right angles to the magnetic field direction should feel no net force, but we already know this can't be so. In the original Stern-Gerlach experiment, we expected that some fraction of the little atomic magnets would happen to have their axes perpendicular to the magnetic field, and so would pass through undeflected, but no such atoms were found: everything has to go one way or the other.

You can guess what actually happens. A beam of up electrons passed through a horizontal Stern-Gerlach magnet will be split into two beams, one to the left and one to the right, and the two beams will be of equal intensity. Each up electron has a fifty-fifty chance of going either way. (See Figure 2.)

And, you can now predict with confidence, if you were to

FIGURE 2

What you might expect (and indeed get . . .)

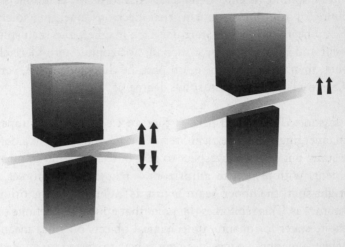

. . . not you what might expect (but get anyway)

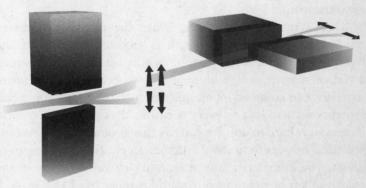

An electron that's already known to be "up" indeed comes out of the "up" path of a second vertical Stern-Gerlach magnet. But if an "up" electron passes through a horizontal Stern-Gerlach magnet, it comes out of either the left or the right path with fifty-fifty probability.

take the left-hand beam of electrons from this magnet, and guide it through another vertical Stern-Gerlach magnet, it would be split again into two equal up and down components (meaning, if you think about it, that electrons emerging solely in the up beam of the first Stern-Gerlach magnet, and then split into left and right beams by a second, horizontal Stern-Gerlach magnet, then seem to acquire, on passage through a third, vertical Stern-Gerlach magnet, some degree of "downness").

The lesson to be learned here is that an electron in the upper beam emerging from a vertical Stern-Gerlach magnet is in a definite state, but only with respect to that particular kind of measurement. With respect to a horizontal magnet, which forces a left-right split, the upper beam is just as indefinite as the original beam was. This reinforces the point that any talk of definite or indefinite states for quantum mechanical objects only has meaning when some specific measurement is being referred to: an electron may be unambiguously "up" with respect to a vertical Stern-Gerlach measurement, but the same electron is in an entirely indeterminate state with respect to a horizontal, or left-right, measurement. Any property must be defined in conjunction with a measurement.

There's another lesson here. When the up beam is passed through the horizontal magnet, it is split equally into left and right beams. Either of these left or right beams, if passed through a vertical magnet, will yield up and down beams of equal intensity. This can go on ad infinitum (although, of course, the beam intensity is cut in half on every occasion, making the experiment increasingly hard to do). We can say this another way: if you use a vertical magnet to find out whether an electron's spin is up or down, you completely lose the ability to say whether it will come out left or right in a horizontal magnet. And vice versa. If you insist on acquiring one piece of information about an electron, you have to settle for ignorance of some other piece of information.

This sounds like something you may have heard of before: an uncertainty principle. If you want to know one thing, you have to forgo knowledge of something else. The uncertainty principle in quantum mechanics is often said to mean that you cannot simultaneously know the position and velocity of a particle; the more accurately you try to measure one, the more uncertainty you have to accept in the other. But that's just one example of how the uncertainty principle operates; or, if you prefer, just one example of a whole constellation of uncertainty principles. Uncertainty principles are a general and inescapable feature of quantum mechanics, and represent one of the key differences between quantum and classical theory, which has no such thing. If, in a series of Stern-Gerlach measurements, you were measuring some intrinsic property of the electron, then classical physics would say that once you had made the first measurement, you would know exactly what state the electron was in, and could predict the result of any future measurement. But in quantum mechanics, to repeat the point once more (it's important!), measurement is an act by which the measurer and the measured interact to produce a result. It's not simply the determination of a preexisting property.

The uncertainty principle is bound up with the probabilistic nature of quantum measurement—that is, the fact that quantum mechanics makes predictions about the probabilities of getting this or that outcome, but can never say with certainty, in any individual case, what is going to happen. You can say that an up electron will definitely come out "up" if you do another "up" measurement, so there's no uncertainty, and no probability involved. Then again, it's also not a very interesting experiment, since you already know the answer and don't learn anything new. If you want to find out something different, by sending your up electron through a horizontal magnet, you have to make do with the probabilistic statement that it has a fifty-fifty chance of going either left or right.

Coin tossing and weather forecasting

Probability, so we are beginning to think, underlies many of the difficulties in grasping quantum mechanics. The fact that we cannot say what any particular electron will do when it passes through a Stern-Gerlach magnet, only that it has a fifty-fifty chance of going one way or the other, is a large part of the reason why we can no longer think of the electron's spin as having, before the measurement is done, any particular value. It's the need to accommodate all possible results of all possible orientations of spin measurement that requires us to think of spin as being indeterminate, or undefined, until a measurement is made.

But probability is not unique to quantum mechanics. The conventional example of a fifty-fifty outcome is tossing a coin—it's almost a definition of what we mean by a fifty-fifty chance, and yet coin tossing does not seem to present us with any great conceptual difficulties, or to have much to do with quantum theory.

And, if we want another example, we can think of the weather forecaster telling us that there is a 50 percent chance of rain tomorrow afternoon. No quantum mechanics there either, we hope; forecasting the weather is hard enough already. What, if any, is the difference between the quantum-mechanical fifty-fifty chance of getting either an "up" or a "down" electron out of a Stern-Gerlach magnet, and the familiar fifty-fifty odds that come up in coin tossing or weather forecasting?

When a coin is spun into the air, its motion obeys the classical laws of mechanics: it spins at a constant rate, following as it does so a parabolic trajectory that takes it up to some maximum height before descending. If you had the resources of a large physics laboratory at your disposal, you might imagine building a machine to predict the outcome of coin tosses. You could use a video camera to record the flight and spin of the coin just after it was tossed into the air; then, digitizing the images, you could feed that information into an appropriately programmed computer that would predict the coin's flight path. If you knew from what height the coin was tossed, and how far it had to drop before hitting the floor, the computer could be programmed to figure out how many times it would spin before landing. Barring a certain amount of uncertainty about the landing (would the coin bounce a little? if it happened to land exactly on edge, which way would it drop?), this machine could make a reasonable job of predicting heads or tails before the coin landed.

In other words, the reason we say a coin has a fifty-fifty chance of landing heads or tails is because we have insufficient information about its flight dynamics to make a reasonable prediction one way or the other. As long as the coin is tossed reasonably high and spun fairly fast, its flight path and spin are complicated enough that it could as easily land on one side or the other, and we are reduced to guessing blindly how it will end up. But if, with a sophisticated laboratory full of fancy equipment, you could observe its motion early on and make a rapid calculation, you would be able to make a good estimate of the outcome. It might not be possible, given the limitations of computer power and the full complexity of the coin's motion, to get the answer right every time, but you should be able to improve the prediction from the simple fifty-fifty guess. On every individual coin toss, your machinery would briefly watch the coin's motion, do its calculation, and predict heads or tails. Perhaps it would get the right answer 80 percent of the time. A better program might improve that to 90 percent.

The weather forecast is a little different. On Friday evening, the forecaster says there is a 50 percent chance it will rain on Saturday afternoon. By the time Saturday afternoon comes around, either it is raining or it is not. But suppose, on Saturday evening, you were able to go back to Friday night, winding the weather patterns back in time exactly as they had unfolded, and thus restore them to their original starting position. If you then allow the clock to move forward once more, so that the events of the previous twenty-four hours are repeated, the result would be precisely the same, as if you were playing the same movie over and over. If it rained on the first Saturday afternoon, it would rain on all the identical repeated Saturday afternoons, and if it didn't, it wouldn't.

When we talk about tossing a coin repeatedly, we are talking about a series of similar but different events: no one is skillful enough to toss a coin with exactly the same force and spin every time, so that it always comes up the same way when it lands. With weather forecasts, on the other hand, we are talking of a unique set of circumstances that develops once, and once only, to a particular result. The weather forecaster's task is something like trying to use the video camera and computer to predict the outcome of a particular coin toss, except that the dynamics of weather are much more complicated and the information on which the prediction is based is much more limited. It is like trying to forecast the outcome of the coin toss from a blurred video image, with only a primitive computer to make the calculation. You would be able to make some sort of prediction, but only an inaccurate one. If you get hold of a better video camera and a faster computer, you should be able to make a much better prediction.

Likewise, if the weather forecasters can gather more detailed information of weather patterns on Friday evening and plug them into a bigger, faster computer with a more elaborate program that takes into account more of the subtleties of atmospheric dynamics, they should be able to provide more accurate forecasts. As their predictive powers improve, they should be able to say that there is a greater than 90 percent chance of rain the next afternoon, or less than a 10 percent chance. If they had perfect knowledge and an unflawed understanding of the weather, they should be able to say with certainty either that it will rain or that it will not rain. With perfect understanding, there would be no element of uncertainty in weather forecasts, and no need for probabilities. Similarly, if the coin-tossing prediction machine were to be built in the most sophisticated supercomputer laboratory, and if the computer programmers were able to include in their program not just the basics of coin flight but all kinds of details of air resistance, the way the coin bounces on different kinds of carpets, and so on, their predictions too would approach perfection, and they would be able to say with certainty that the coin would land on one side or the other.

In all such "classical" applications, probabilities are a cover for ignorance. We say that a coin has a fifty-fifty chance of coming up heads, or that there's a fifty-fifty chance of rain tomorrow afternoon, because we don't have enough information to make any better a prediction. In principle, though, we can imagine acquiring more data and a more sophisticated understanding of the physics involved so that in either case we could hope to make steadily more accurate predictions.

Quantum mechanics is different. When the theory says that an electron emerging from a Stern-Gerlach magnet has a 50 percent chance of coming out in the up beam and a 50 percent chance of coming out in the down beam, that's all there is to be said. No more information is available, even in principle. No additional data could be gathered that would enable the experimenter to say, of any individual electron passing through the magnet, that it will go one way or the other. Predictions in quantum mechanics are probabilistic not because of insufficient information or understanding, but because the theory itself has nothing more to say. The statement that an emerging particle has a fifty-fifty chance of being in one state or the other is all there is to say.

This is what Einstein found so much to his distaste. He could accept the idea of classical probability as a way of admitting that we have incomplete information about a system, but he could not bear the idea that you could have complete information about a system and still have to make do with a statement of probabilities as to the outcome of an experiment. Thinking this way, Einstein maintained that quantum mechanics must itself be incomplete, and that there must be a deeper, more detailed theory that would include all the necessary information to enable a physicist to make full and certain predictions of the outcomes of experiments, not mere statements of possibilities and probabilities.

Einstein, and a few others, believed that a more complete theory of this sort was essential to make physics whole again, by removing the stain of probability from its core. Bohr, along with the majority of physicists, came to the conclusion that looking for such a theory was a misguided, quixotic venture, motivated by romantic thoughts of what physics ought to be rather than by a pragmatic understanding of what physics is.

Not just electrons

If quantum mechanics had to do only with the magnetic properties of atoms and electrons, as revealed by specially constructed magnets, it would be an interesting theory, full of conceptual novelties, but not a revolutionary new version of fundamental physics. But it turns up everywhere. Here's another example of the same kind of probabilistic measurement turning up in a quite different place.

Light, as James Clerk Maxwell powerfully demonstrated in the middle of the nineteenth century, is a wave motion, and a cousin to all other kinds of electromagnetic radiation—radio waves, infrared and ultraviolet, X rays, and so on. The wave nature of light had been demonstrated in all kinds of experiments through the preceding centuries, but Maxwell tied it all together.

One of the properties of light is polarization, which can easily be visualized if you think of wave motions on a clothesline or a skipping rope. A vertical wave motion up and down is one kind of polarization; a horizontal wave motion back and forth is another. You can also have circular polarization, where you move the rope's end rapidly in a circle with your hand, creating a sort of spiral wave that travels down the line.

Polarized sunglasses block out some kinds of polarization and allow others through, a useful thing because light reflected from the surface of a lake or glancing off car windshields becomes polarized in one direction, so that if you can block that with your sunglasses, you will cut out most of the reflected glare while leaving intact light from other sources. To pursue the clothesline analogy, you can think of a fence of vertical palings as a polarizing filter; if the clothesline passes between two of the slats in the fence, up-and-down waves (vertical polarization) will pass easily through, but side-to-side waves (horizontal polarization) will not. If the wave motion is at some angle between vertical and horizontal, a proportion of the wave motion will pass through, emerging on the other side of the fence as strictly up-and-down motion, while the remainder will be blocked, and will bounce off instead of passing through.

The same holds true for polarized light and polarizing filters. You can make vertically polarized light by passing any old kind of light through a vertically aligned filter; what emerges from the other side will be strictly vertically polarized. That light will pass unobstructed through another vertically polarized filter, but it will be completely blocked by a horizontally aligned filter. A filter set at, say, forty-five degrees to vertical will let through half the vertically polarized light, and block the rest.

Easy enough. As long as light is a wave motion, its intensity can be divided and subdivided as much as you like, so you can pass a beam of light through a whole series of polarizing filters, cutting out a little bit here, a little bit there. The intensity will diminish at every step, but that's no problem (theoretically, anyway; any such light beam will eventually become undetectably faint if you mess around with it enough).

But now comes the new quantum mechanical ingredient. Light has wave properties, true enough, but according to quantum theory light energy comes in little packets, called photons, which cannot be subdivided. A typical light beam contains countless photons, which individually carry only a tiny amount of energy, and for most practical purposes it's immaterial that it consists of lots of little indivisible bits, and is not a continuous stream.

But you can see that if we take light to consist of photons, we immediately run into a conceptual problem: what happens when a single photon of polarized light encounters a filter? If it's a vertically polarized photon, and it meets a vertically aligned filter, then there's no problem—the photon goes through. And if it meets a horizontally aligned filter, it gets blocked or reflected—also no problem. But if the filter is at forty-five degrees to vertical, then we are in trouble. Classical wave theory would say that half the light energy goes through, and half gets reflected. But if, as quantum theory dictates, the photon cannot be divided in two, then the entire little packet of energy must either go through or get reflected. It can't do both, and it can't split up, so it has to do one or the other. (See Figure 3.)

And because, if you send a whole stream of photons through this filter, half would have to go through while half were reflected, it has to be the case that any individual photon has a fifty-fifty chance of doing one thing or the other. Once you accept the idea of light as photons, probability cannot be avoided. And as with electrons passing through a Stern-Gerlach magnet, this is the kind of probability where it's not a matter of

FIGURE 3

Classical waves get split into two . . .

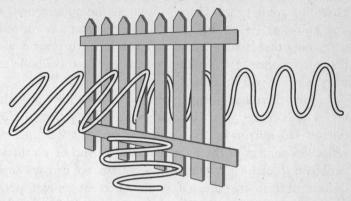

. . . but quantum waves have to go one way or the other.

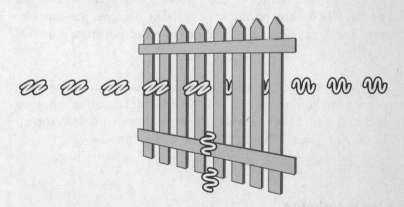

Classically, a continuous wave with some angle of polarization is divided smoothly into a vertically polarized wave that passes through the filter, and a horizontally polarized wave that bounces off. But quantum mechanically, the photons cannot be divided, and their behavior cannot be individually predicted; each one must either pass through or bounce off the polarizer, with the appropriate probabilities.

being able to get a more precise answer if you could find out more about the individual photons. There's nothing more you could possibly find out that would tell you what any particular photon is going to do. Once you know the polarization, then you know, as far as this particular experiment is concerned, everything that matters. And still you have to put up with a fifty-fifty chance of either outcome as the best available statement of what will happen.

In fact, there's an even simpler way to see how probability enters into things. Just think of light being reflected from a mirror. No mirror is perfect, but a pretty good one might reflect 95 percent of the light that hits it. And here's the same problem: if lots of photons hit the mirror, we can say that 95 percent of them are reflected, while 5 percent get lost (perhaps passing through the reflecting surface, or being absorbed by it). But for any single photon that hits the mirror surface, the best we can say is that it has a 95 percent chance of being reflected, and a 5 percent chance of getting lost. Once again, as soon as we think of light as photons, probability becomes inescapable, even for so familiar an occurrence as reflection from a household mirror.

Oh, but we forgot something. Whoever said that light has to consist of photons, and why? If we could forget about photons and stick with light as a wave, these problems wouldn't crop up in the first place. Why do we have to have photons?

Enter the photon

If a case is to be made for the existence of little packets of light energy by the name of photons, it has to rest on experimental evidence. How did photons enter the minds of physicists?

Photons were in fact among the very first offspring of quantum theory, though they suffered through an uncertain childhood during which physicists were not quite sure whether to allow them an independent existence or not. They came about, at the turn of the nineteenth century, as a consequence of the German physicist Max Planck's solution of a difficult puzzle presented by classical physics.

Conceptually, the puzzle in question should have been easy. Physicists knew by this time that the heat energy of material substances was a manifestation of internal atomic motion, so that the faster the little atoms jiggled around the hotter a body was. Temperature is nothing but the average speed of the atoms: faster equals hotter.

At the same time, physicists also knew that electromagnetic radiation—radio waves, infrared, light, whatever—carried energy. This was a continuous form of energy, characterized by the average amount of electromagnetic wave motion in some volume of space rather than the average energy carried by a discrete and identifiable atom, but energy is energy no matter what is carrying it, atoms or waves, so there ought to be some correspondence between the two.

In particular, physicists set themselves to solving the idealized problem of an enclosed empty box, maintained at some fixed temperature and containing electromagnetic radiation. There had to be some sort of equilibrium: if there were more energy in the walls of the box than in the enclosed radiation, then energy would move from the walls to the interior, increasing the density of radiation within the box; if there were more energy in the electromagnetic radiation than in the walls, then it would heat up the walls to restore equality. In short, both the walls of the box and the electromagnetic waves inside it ought to carry some comparable amount of energy, and should be characterized by the same temperature, otherwise heat would flow from the hotter to the cooler party until equality was established.

The problem was this: although physicists knew how to calculate the energy carried by an individual electromagnetic wave, they couldn't figure out how to calculate a meaningful temperature for the mixture of electromagnetic waves that would fill such a box.

Here's what happened when they tried to work out the temperature of a collection of electromagnetic waves. First, they realized that sustained electromagnetic oscillations in the box could not have any wavelength they chose. Like the vibrations of a violin string, the oscillations had to fit into the available space. A violin's lowest frequency note obtains when the whole string moves up and down together, the ends remaining fixed; the next frequency—twice the frequency, in fact—occurs when the string has a double oscillation, one half moving up while the other half moves down, with the midpoint remaining stationary; then you get a three-part oscillation for three times the basic frequency, and so on.

Similarly, electromagnetic waves in a box would start at the low end with a single oscillation fitting right across the inside of the box. Then there would be twice the frequency, with half the wavelength, so that two whole oscillations would fit inside the box. And so on, upwards in frequency.

Next, standard electromagnetic theory tells you (if you're a physicist at the end of the nineteenth century) that any wave carries an energy proportional to its frequency and to its amplitude (that is, the magnitude of the up-and-down motion). Moreover, you realize that although the oscillations that can fit into the box have to have certain specific frequencies, starting with what musicians would call a fundamental tone, and proceeding through the second harmonic (twice the frequency, half the wavelength), the third harmonic, the fourth, and so on, it seems that this progression goes on to infinity. There's a millionth harmonic, a million-and-first, a billionth, a zillionth. In short, the box is able to accommodate an infinite number of electromagnetic oscillations, all carrying some share of the overall energy.

This is where the problems begin. In the material of the box itself, heat energy is equivalent to the motion of the constituent atoms, of which there is a large but finite number. Some fixed total amount of energy is shared by all these atoms. Some travel more slowly, some more quickly, but there is a well-defined average energy per atom, and this average is what defines the temperature of the box. The more energy you pump into it, the more energy there is to share among the atoms, and the faster, on average, they move. The temperature goes up.

But this simple reasoning cannot be transferred to the electromagnetic waves in the box, because of the infinite number of available oscillations. Physicists could not find any way to figure out how a fixed amount of energy would be shared among these infinite possibilities in such a way as to arrive at a meaningful "average" energy per wave, which could be thought of as the temperature. It's not possible to divvy out the energy equally between all possible waves, because then you would get an infinitesimally small quantity of energy in each one of the infinite number of possible waves. Alternatively, if you try to give each little wave a certain amount of energy, you need an infinite amount of energy overall to give every wave a more than negligible amount. No matter how they calculated it back and forth, physicists could not deal with this problem of infinity. And this meant they could not understand, theoretically, how to define the electromagnetic radiation that would be contained in a box at constant temperature.

Nature, of course, was ahead of the physicists, and solved the problem elegantly. If you make a box such as this, and drill a tiny hole in it to peek inside and see the radiation, you find a perfectly straightforward answer. Experimental physicists at the end of the nineteenth century could perform such a test, and they found, to no one's great surprise, that a heated box contained radiation that went, as they raised the temperature, from a cool orange-red glow to a brighter yellow to a hot blue-white emission. This ascent through the colors of the spectrum, familiar to blacksmiths for centuries, was an ascent through electromagnetic radiation of increasing frequency.

Nature, evidently, knew how to divide up electromagnetic energy so that at any given temperature, the bulk of the energy was in oscillations of a frequency proportional to the temperature: red, then yellow, then blue. Why could theoretical physicists not describe mathematically how this happened?

Max Planck found a way out of the dilemma. What would happen, he imagined, if each individual electromagnetic wave could not carry any arbitrarily small amount of energy (which was in essence the sticking point with all the previous efforts to solve the problem) but had to carry some minimum amount proportional to its frequency, or else none at all? More precisely, Planck suggested that each electromagnetic wave could carry energy only in multiples of this basic amount, so that the energy in any individual wave was a whole number times the frequency of that wave, multiplied by a conversion factor that came to be known as Planck's constant.

How did this solve the problem? Simply because for waves at very high frequency (the zillionth harmonic and more of the fundamental frequency) the minimum unit of energy became so large that it exceeded all the energy in the heated box, which meant that very high frequency oscillations never arose. Planck's "quantization" of the energy in electromagnetic waves—that is, his stipulation that each wave must carry multiples of some basic quantity of energy—meant that the available number of oscillations in a box became finite rather than infinite, because all those oscillations above a certain frequency could be ignored. With a finite rather than an infinite number of objects to carry the energy, the calculation became much like the familiar divvying up of energy between the finite number of atoms in the box. Physicists were then able to come up with a sensible answer for the way energy was distributed among the electromagnetic oscillations, and from that could straightforwardly define an average energy-per-oscillation that made sense. And this average was the temperature of the radiation.

Now the problem was solved. There was a temperature for the atoms constituting the walls of the box, a temperature for the electromagnetic radiation inside the box, and some straightforward arguments showed that these two temperatures would indeed be the same. Equilibrium was established between matter and radiation, and physicists were content at last.

Except for one thing. Planck had decided that oscillations could not carry whatever energy a physicist pleased, but were restricted to multiples of some basic value. This unit, this quantum of energy, this division of energy into little packets, was a new idea in physics. And so the photon was born.

So photons are really real, then?

Well, it wasn't all that clear at the time. The meaning of Planck's ingenious solution to the radiation problem remained controversial for some time. Planck himself didn't want to believe that electromagnetic radiation was fundamentally constrained in this new way, and hoped to find some overlooked subtlety of classical physics that would explain why waves had to carry energy in discrete quantities. Just as the geometry of the box allowed waves to exist only at harmonics of some fundamental frequency, so Planck thought some similar argument to explain the discretization of the energy in those waves would eventually be found. In short, Planck did not by any means like the implication that he had found a new and distinctly non-classical feature of physics, and he never attributed to the little bundles of electromagnetic energy any genuine physical reality. It was just a matter of time before an explanation of the rule he had devised came along.

In a conceptual or philosophical way, physicists could argue the question endlessly one way or the other. Are photons real, or are they mathematical constructs hiding some unknown physical principle? But as time went by, it became apparent that certain experimental results were easily and consistently understood on the assumption that photons were genuine and universal physical objects. Photons kept turning up all over. Two pieces of evidence in particular are worth mentioning.

First, there's the photoelectric effect. Certain metals emit electrons when light is shone on them. It's not too hard to see why electromagnetic energy impinging on a metal surface could shake loose a few electrons. The reason metals conduct electricity is that some of their electrons are free to roam throughout the bulk of the material, hopping from atom to atom, so it's an established fact that electrons in metals are not as securely tied down as they are in other substances. And you can imagine that throwing energy at a metal, in one form or another, might knock out electrons from time to time. But two details of the photoelectric effect eluded easy explanation.

Experimenters had found that for electrons to be liberated from any particular metal, the light raining down on the surface had to be of some minimum frequency, which depended on the metal in question. Green light, for example, will expel electrons from a piece of sodium metal, but to knock electrons out of more common metals, such as copper or aluminum, you need to go to more energetic ultraviolet light. Moreover, it was found that, once electron liberation has begun, turning up the intensity of the light increases the number but not the energy of electrons that are popped out, while turning up the frequency of the light brings out electrons of higher individual energy, but at the same rate as before.

These facts are hard to understand using a wave theory of light, in which the energy carried by waves is a product of the frequency and the intensity: low frequency radiation of high intensity is the same, in terms of the energy conveyed to the surface of a photoelectric metal, as higher frequency radiation of lower intensity, and there's no reason why frequency and intensity should have such markedly different effects.

But then Albert Einstein, working with the idea that light is a beam of photons, explained the photoelectric effect very simply. Imagine that a photon bangs into a metal, and has to eject an electron. If the electron is bound into the structure of the metal with some attractive force, it will take a certain minimum amount of energy to knock it out. Since photons carry energy in proportion to their frequency, it follows that a photon capable of knocking out an electron must have at least that energy, which means its frequency must exceed some threshold value. (Two photons of lower frequency, or energy, might in principle be able to eject a single electron, but they would have to hit the electron one right after the other, and the necessary precision of timing makes this an unlikely eventuality.) This explains why the light has to exceed some minimum frequency before it can liberate electrons from a metal. And because different metals have different atomic properties, that minimum frequency will be a characteristic property of each metal.

Turning up the intensity of the light beam amounts to bombarding the metal surface with more photons. Individually, the photons knock out electrons as before, but since there are more photons, the light beam as a whole generates electrons faster. On the other hand, turning up the frequency of the light amounts to sending in photons with greater individual energy. Each one is able to knock out only a single electron, but because the photon plunges in with more energy, the electron receives a bigger push when it is knocked out. This explains why a more intense beam generates more electrons, but a higher frequency light beam generates more energetic electrons.

The photon theory of light can therefore explain very simply experimental facts at which the wave theory balked. Einstein received his 1921 Nobel Prize for this explanation of the photoelectric effect, and it's often suggested this was because the Nobel committee chickened out of giving him a prize for his more fundamental and controversial theories of special and general relativity. On the other hand, establishing the reality of photons, which was just as controversial a matter at the time, is hardly a throwaway achievement.

•

An even more direct demonstration of photons acting like little particles came from the American physicist Arthur Compton in 1922. It had been observed that X rays bounced off certain crystals came out with less energy than they took in, and that the amount of energy they lost depended on the angle they were reflected through. Compton concluded that this could be understood if you thought of the X rays as individual photons bouncing off electrons belonging to atoms in the crystal. If you think of both the electron and the photon as particles, this is strictly a billiard ball collision, and there is a direct relationship between the direction the electron bounces off in and the energy it has (compare a billiard ball banging directly into a stationary one, and sending it off at some speed, with a glancing collision that just sets it gently rolling to the side). Moreover, the photon's energy is also changed in the collision, and because, according to Planck's hypothesis, energy and frequency are proportional, the before-and-after change of photon energy amounts to a change of frequency that can be fairly easily detected.

Compton demonstrated that the mechanics of such collisions were just what you would expect from a billiard ball model of photon and electron. With a wave theory of light, by contrast, there's no reason why the photon's frequency would be changed by an interaction with an electron. For this work Compton too received a Nobel Prize, in 1927.

This is by no means the only evidence that addresses the reality or otherwise of the photon, but these two long-standing arguments are as good as any to demonstrate that photons are not mathematical devices, as their inventor Planck wished it, but real little particles.

Except, alas, it's not that simple. Photons may be little packets of energy, little particles of light, but light incontrovertibly has wave properties too. It's a wave; it's a particle; it's a wave *and* a particle.

Particle or wave?

Long before quantum mechanics came along, when Isaac Newton and his peers and successors were building the foundations of modern physics, there was a dispute as to the nature of light. Newton, among others, favored the idea of "corpuscles"—for one thing, the simple fact that light travels in straight lines immediately brought to mind the notion of a stream of little projectiles. What exactly those projectiles might be, neither Newton nor anyone else could guess.

But other evidence suggested that light was a wave motion. Here's one of the fundamental experiments to support that view.

In 1801, the English physicist Thomas Young shone light of a single wavelength, from a single source, onto a screen with two narrow slits. Light passed through the slits, of course, and Young placed another screen a little way beyond the first so that he could see the pattern of light that emerged.

If light were "corpuscles" (particles, bullets, whatever), then these corpuscles would pass through each of the slits, continue on in straight line trajectories to the second screen, and make two bright patches. That isn't what Young found.

What would happen if light were a wave motion? Light passing through the narrow slits should spread out on the other side of the screen, not continuing just in a straight line but diverging sideways too, as ocean waves entering a harbor spread outwards after passing through a gap in the harbor wall. On the far side of the screen, the wave motions of light spreading from the two slits will merge together, so that what you see on the second screen is a combination of waves coming from both slits.

Now, it's important at this point that the light is all of the same wavelength, and that it all came originally from the same source. What this means is that the light everywhere is coherent: the waves emerging from both slits march in unison, moving up and down to the same beat. If Young had put a separate light source behind each slit, there would be no harmony between the wave motions emerging from the two slits. (See Figure 4.)

Look now at the middle of the second screen, at the spot that's precisely equidistant from the two slits. Because the light waves from the two slits are coherent, and because they have to travel the same distance to get to this spot, they arrive together: the light waves beat up and down as one, and you see a bright spot.

Cast your gaze now a little to the side, to a spot where the distance to one slit is exactly half a wavelength of light greater than to the other slit. Because of the difference in distance, when light waves from one of the slits reach this spot at the peak of their motion, waves from the other will be at their trough. Light from the two slits is out of phase, as the physicists would say, and the waves cancel each other out. At this spot, therefore, darkness.

And so on. Look at a spot the same distance again to the side, to which the difference in distance from the two slits is one full wavelength of light. Here, the wave motions will once more arrive together. They will be in phase, and the screen will be bright.

If light is a wave, then on the farther screen in Young's experiment there should appear a pattern of bright and dark stripes extending to both sides, bright where the two wave motions add together in phase, dark where they cancel each other out. Notably, there will be a bright spot in the middle, which in the corpuscle picture of light would be in darkness.

FIGURE 4

Interference between waves

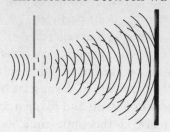

No interference between particles

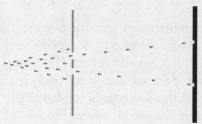

The quantum version

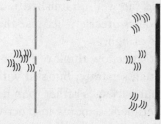

Waves passing through a double slit are divided into two separate but related wave streams, which interfere to create a pattern of stripes. Bullets passing through two slits carry on independently in straight lines, with no interference. Quantum mechanically, the continuous classical wave is divided into bulletlike photons, yet wavelike interference still occurs.

What results is called an interference pattern, for the simple reason that it arises from the interference of one wave motion with another, and it's just what Young saw in 1801. Light consists of waves, therefore, not of particles.

(And it's worth observing that regardless of the details of the interference pattern produced in this experiment, it's difficult by any means to see how two beams of light could interact to produce darkness, if light consists of corpuscles or particles. Particles are either present or absent, and two particles of light arriving at the same place at the same time would always add together to manifest themselves as more light. How could particles of light cancel each other out?)

But now, bringing ourselves back into the twentieth century, we have to wrestle again with the idea of "particles" of light—photons. Except that we can't entirely regard photons as bullets, in the way that Newton perhaps imagined, because photons are imbued with some of the wave properties of light too. They have wavelength and frequency. Even though Max Planck invented photons to explain how radiation of different colors— red to yellow to blue—emerges from a heated object, his explanation had to include the well-established fact that we recognize those different colors precisely because they correspond to different wavelengths of light.

Evidently we must somehow think of photons as little packets of waves. Individual, separate, but still with wavelike character; fuzzy, spread-out packages rather than hard little kernels. Young's two-slit experiment remains instructive if we want to understand better what photons really are.

Let's set up the two-slit experiment once again, but now that we are in the twentieth century let's suppose we can use a laser to generate the light that's sent to the double slits. A laser gives us better control of the light, and we can think of it as generating a stream of photons all of the same wavelength, frequency, and energy, and at some rate we can carefully set. Switching on the laser, we see the stripes, and perhaps we think of the stream of photons, trillion upon trillion, as somehow combining together to create what classical physicists would have imagined as a coherent, uniform wave motion.

If we turn down the intensity of the laser, the pattern of stripes gets fainter. No surprise there. In the classical view, the intensity of light corresponds to the height of the wave motion; as the intensity is reduced, the wave motion gets shallower, but it remains a wave, and the striped pattern due to interference between the waves from the two slits is unaffected except in its intensity. You can turn down the laser as much as you like, and still see the stripes, provided your eye, or whatever fancy device you are using to record the images, is sufficiently sensitive.

But what if we insist on thinking of light as photons? They're some kind of particle, if not exactly bullets, and didn't we decide a few paragraphs ago that light-as-particles would not produce an interference pattern?

And if we turn down the laser to such a low level that it generates, let us imagine, one photon every three seconds, is that photon not obliged to go through one slit or the other? In which case, should not the pattern of stripes disappear, since interference, which produces the stripes, is by definition the interaction between the light emerging from the two slits? We seem to have run into a dilemma. The existence of the striped pattern of light and dark at the screen seems to demand that light from the original source must be divided into two fractions, which emerge from the two slits and interfere with each other on the other side. Yet if the light wave is actually a stream of photons, and if the intensity of the light is so low that photons proceed through the apparatus one by one, it seems that each must go through one slit or the other; there can then be no interference because a single photon emerging from a single slit has nothing to interfere with.

The incontrovertible fact is that you can perform Young's two-slit experiment with a laser, and you can turn down the laser as much as you like, so that it doesn't send out another photon until the previous one has had ample time to pass all the way through the apparatus, and yet you still see the interference pattern. How does this square with the idea of light as particles?

One photon at a time

You may well be wondering what it means to say that a single photon can produce an interference pattern in the two-slit experiment. A single photon, when it strikes the screen, can't be everywhere at once, and generate a whole pattern of bright and dark stripes all by itself: it has to produce a little bright dot one place or another.

This is indeed the case. Let's describe what's happening more carefully. You can set up a two-slit experiment using a very low intensity laser that, to all intents and purposes, generates photons one by one. Each of these photons passes through the apparatus, strikes the screen, and makes a little mark. We could imagine that the screen is a photographic film, so that each photon striking the emulsion leaves a blip that we can later develop to create an image. But since we have more modern equipment at our imaginary disposal, let's suppose that the screen is covered by lots of little electronic detectors that can record the impact of a photon. The tiny detectors send their signals to a computer, which records them and creates, on a monitor, an image that builds up from the accumulated impacts of successive photons on the screen. We switch on the experiment and watch as the impact of each photon marks the screen. What kind of image will be built up, as we wait and watch?

At first, there will be little dots here and there, but as the photon impacts accumulate, a pattern begins to stand out: we see the stripes of the interference pattern. If the intensity of the laser was turned up so that a million photons went through the apparatus all at once, we would see the entire pattern emerge in one piece. But with the laser intensity turned down low, we see the same pattern arise dot by dot. This is what we mean when we say that individual photons passing through the two-slit apparatus generate an interference pattern.

And what this also means is that probability enters the picture again. In the simple case of light reflecting off an imperfect mirror, we saw that you must resort to probability: 95 percent of the photons are reflected, the remainder are lost, and for any individual photon the best you can do is allot the appropriate odds.

The same thing now happens in the two-slit experiment, when you think about it from the point of view of individual photons. Collectively, the photons strike the screen so as to build up the requisite pattern of stripes. Individually, you can never say where any particular photon is going to land, but you can say that it has a relatively high probability of hitting the screen at a place where a bright stripe is going to emerge, and a relatively low probability of hitting the screen in one of the dark stripes (right at the very center of a dark stripe, the probability will fall to zero, and no photon will hit the screen there).

This is more complicated than the case of reflection, where a photon has to go one way or the other, with ninety-five to five odds, or for that matter the Stern-Gerlach magnet, where the electron has a fifty-fifty chance of coming out in either the up or the down beam. Here, the probability according to which a photon lands on the screen must be high, low, high, low, high, low, and so on, to add up to a pattern of stripes.

To describe this kind of situation mathematically, physicists construct a formula that gives the pattern of probabilities; its value goes up and down across the screen, so that if you want to know the chance of a photon hitting the screen in one place, the value of this mathematical expression at that place gives you the answer you want. This formula is what physicists call the wavefunction for the photon, for this particular experiment.

(In fact, the wavefunction, properly defined, is something like the square root of the required probability, but for the time being let's just say that the wavefunction is a mathematical device that allows you to figure out the correct probability for the photon hitting the screen at any place you choose.)

The wavefunction is a ubiquitous but slippery concept in quantum mechanics. Wherever you meet probabilities, you'll find a wavefunction. For the photon reflecting off an imperfect mirror, the wavefunction is (translated from the appropriate mathematical terminology) "95 percent reflected, 5 percent lost." For the electron emerging from a Stern-Gerlach magnet, the wavefunction would be "50 percent up, 50 percent down" or, if the Stern-Gerlach magnet is horizontal rather than vertical, "50 percent left, 50 percent right." That's an important point: the wavefunction is not a property of the electron in isolation, but of the electron and whatever measurement is being made. Different measurements, different Stern-Gerlach magnets, require different wavefunctions.

The results of successive Stern-Gerlach measurements can be easily translated into the language of wavefunctions. If we send an electron toward a vertical Stern-Gerlach magnet, knowing nothing whatever about the electron, then we would describe its wavefunction as "half-up, half-down" (using "half" to mean "50 percent," for brevity), meaning that either outcome is equally likely. What happens if the electron goes through the magnet, and emerges in the up beam on the other side? If we were to send it through another vertical Stern-Gerlach magnet, we know for sure it would come out in the up beam again, so we would have to describe its wavefunction as simply "up."

But if we took this up electron and sent it through a horizontal Stern-Gerlach magnet we would find, as we discovered before, that it has equal chances of coming out left or right. And so we would have to describe its wavefunction as "half-left, half-right." And if the electron then emerged in the left beam, we could describe it at that point by a definite "left" wavefunction, except that if we sent it through another vertical magnet, we would have to describe it instead as "half-up, half-down."

In other words, to reinforce the point we already made, an electron by itself is not described by one unique wavefunction; the way you describe it, the wavefunction you use, depends on what you plan to measure. And although the wavefunction obviously depends on the state of the electron, and on what you know about it, it can be misleading to think that the wavefunction somehow "is" the electron. It's better to say that a wavefunction describes a system—the thing being measured and the measurement being made—rather than being an independent description only of the thing being measured.

You can see this distinction also in the wavefunction account of the two-slit experiment. We covered the screen with little electron detectors, which are small but of some finite size. Think of them as little boxes into which a photon might fall, as in the carnival game where you have to throw Ping-Pong balls into an array of glass jars set out on a table. The wavefunction you would use in this case would depend on the size and placement of the detectors, since the only thing you measure is "did the photon go in this detector or that?" The wavefunction might be described as "1 percent detector A, ½ percent detector B, ¼ percent detector C. . . ."

But if your physics laboratory were less lavishly equipped, you might have to make do with cheaper, bigger photon detectors, and fewer of them. The boxes into which each photon might fall would be bigger, and the probabilities correspondingly larger. The wavefunction might now be "2 percent detector X, 1.5 percent detector Y, 1 percent detector Z. . . ."

And if you went to an extreme where, for some reason, the only thing you cared about was whether the photon ended up to the left or the right of the center line, the wavefunction would be "half-left, half-right" just as in a Stern-Gerlach measurement.

The most elementary way to think about wavefunctions, in short, is that they encapsulate the information you need to know in order to figure out the probabilities of different possible outcomes of a given experiment or measurement. This is a cautious definition, but caution, it turns out, will keep us out of a lot of trouble.

Learning to live with uncertainty

We already met an example of the uncertainty principle, in the case of an electron going through successive Stern-Gerlach magnets. An electron emerging from a vertical magnet in a definite "up" state was nevertheless in an indefinite "half-left, half-right" state with respect to a horizontal magnet, and vice versa. You can be certain about one aspect of the electron's characteristics, but you pay for it with uncertainty about another.

The archetypal demonstration of the uncertainty principle, and the one used by the German physicist Werner Heisenberg in 1927 when he came up with the idea, involves electrons colliding with photons. In the Compton effect, remember, an electron and a photon collided in billiard-ball-like manner, with the photon's frequency changing in proportion to its change of energy, an effect that demonstrated the reality of photons.

We still haven't got very far in deciding what a photon really is, but at least we know it's not exactly like a billiard ball. It has to be some sort of fuzzy little wave package. Heisenberg showed that the fuzziness has pervasive and inescapable consequences.

Suppose you want to figure out where an electron is, by bouncing photons off it. If you direct your photons toward a place where you think an electron is hiding, and see a photon bounce off in some different direction, and with some different energy, just as in the Compton effect, you can try to reconstruct the collision to see where the electron was when you hit it. You'll also be able to infer how fast the electron was moving because (if you momentarily bring back to mind the idea of two billiard balls) you can see that the way a photon and an electron collide and bounce off each other depends on the position and motion of both objects.

By now we're getting used to the idea that when photons do things, probability crops up. The same is true here. Going back temporarily to a wave picture, you can imagine the interaction of an electron with light as something like the interaction of ocean waves with a rock sticking out of the water: the rock creates a disturbance and the waves ripple off in all directions. Similarly, light bouncing off an electron would create a pattern of disturbed light spreading out from the electron in all directions. Now switch again to thinking of light as photons, and a familiar problem arises. If photons, collectively, have to scatter off the electron in such a way as to create the appropriate geometric pattern, then photons, individually, must be assigned different probabilities of scattering off in different directions so as to build up the correct overall pattern.

In an individual encounter between one electron and one photon, therefore, you can't say with certainty that the photon will bounce off in this direction, with this amount of energy. You can only say that it has some probability of behaving in such a way. And to describe the probability pattern for the scattering of the photon, we use, of course, a wavefunction. The wavefunction for Compton scattering contains the probability for the photon to emerge from its encounter with the electron in any direction you want to know.

But now if the photon is bouncing off in some direction, the electron must suffer a corresponding recoil. Physicists think of recoil in terms of a particle's momentum, which is speed multiplied by mass, because the laws of dynamics are such that if one particle goes off in one direction with a certain amount of momentum, the other particle must carry the same amount of momentum away in the other direction. Momentum is conserved (you can't lose any of it or come out with more than you took in), so momentum rather than speed is the important quantity. At any rate, if the scattering of the photon is described by a wavefunction that summarizes the appropriate range of probabilities for its position and momentum, then so too the electron's position and momentum has to be described by another wavefunction to cover the range of possible recoiling directions.

And now we see that if we want to work out where an electron is by bouncing a photon off it, we have to make do with probability, not certainty. Just as the prediction of where the photon will end up can be described only in probabilities, using the appropriate wavefunction, so the inference of where the electron was, and how fast it was moving, when the photon hit it, must also be described by probabilities. A photon emerging from the encounter in some direction and with some momentum might have emerged from any number of possible encounters. You can't say there was one specific collision that must have occurred to generate the photon you measure.

As Heisenberg realized, we are not entirely at the mercy of nature. We can choose freely the energy of the photon we send in, and the information we can infer will change correspondingly. In a "soft" collision between an electron and a low-energy photon, the photon can bounce off with similar probability in almost any direction, and the electron doesn't recoil very much, whereas in a "hard" collision, an energetic photon is more likely to shove the electron out of the way and carry on more or less the same direction.

The implication is that by using an energetic photon we can get a better idea of where the electron is, but have little idea of how fast it was moving, whereas a soft collision gives us a poor idea of the electron's position (because the collision geometry is so uniform over all directions), but a reasonable idea of its momentum. This is the classic statement of the uncertainty principle: you measure the properties of an electron so as to obtain an accurate idea of its position but not of its momentum, or you can measure in a different way to obtain a good estimate of its momentum but poor knowledge of its position.

Because the probabilistic nature of the electron-photon encounter is intrinsic to quantum mechanical theory, and cannot be evaded by more ingenious experiments or more careful detection, you can never escape these constraints. The uncertainty principle is a part of nature, not a consequence of our technical limitations. Just as in the case of successive magnetic measurements of an electron, where you can have up-down or left-right information but not both, here you can have position or momentum information, but not both. And the cause is the same: probability. The inability to predict, except in terms of probabilities, the outcomes of different kinds of measurements leads inescapably to an inability to obtain at the same time all the information about an object that you might want to know.

Because of all this, any description of the electron's position and motion has to be in terms of a wavefunction. Having bounced a photon off the electron and obtained, by inference from the photon's properties, estimates for the probability of the electron being in such-and-such a position and having such-and-such a momentum, we have in effect constructed a wavefunction for the electron—a mathematical expression, that is, telling us what chance the electron would have of being in such a place and having such a momentum were we to try and measure those things.

There's a strong urge to say that the wavefunction so obtained represents the electron's position and momentum, and therefore that the electron, loosely speaking, is spread out over space (and in velocity) with some distribution of probabilities. It might be here, it might be there, but it's not really in any one place.

There's some truth in this, in the sense that the wavefunction gives us information that we could use to predict (in the appropriate probabilistic way) the outcomes of further measurements. But remember that if we had done our original measurement differently, using a photon of greater or lesser energy, the wavefunction we would have inferred for the electron would have been different. We might have chosen to get more precise positional information at the expense of greater ignorance of the electron's momentum, and the wavefunction for the electron would have reflected this compromise on our part. As always, what goes into the wavefunction depends not just on the object concerned, but on what we did to it and what we plan to do next.

Retaining our cautious frame of mind, let us say that the electron's wavefunction, as inferred from our measurement, contains the information we could hope to obtain about the electron's position and momentum, given what we already know (given, that is, the results of the earlier experiment). To say that the wavefunction represents, in some hard-to-define way, the actually spread-out position and momentum of a real electron, is something of a leap of the imagination.

To find out where an electron is we must make a measurement, and to predict the result of that measurement we employ the appropriate wavefunction for those circumstances. Wavefunctions are what we use to predict the results of measurements, and measurements are the way we build up knowledge of the world. To suggest that electrons really are, or have some probability of being, in this place or that, in the absence of any means of making such a measurement, is according to our cautious way of thinking a speculation. We can have no knowledge of what an electron is "really" doing when we're not looking at it; we can only make measurements. Anything beyond that is guesswork—which we may, perhaps, be able to verify or refute by means of further measurements.

Is it or isn't it?

If, as a practical matter, we are obliged to describe electrons in terms of wavefunctions, does that mean they also have wavelike aspects that can be experimentally brought out into the open? Of course!

In 1927 Clinton Davisson and Lester Germer in the United States, and also George Thomson in England (whose father, J. J., originally discovered the electron back in 1897), performed experiments that demonstrated the wave character of electrons. They directed beams of electrons, with carefully controlled energy, towards crystals of nickel; the individual atoms of nickel scatter the incoming electrons in all directions, and because, in the crystal lattice, all the atoms are arranged with geometric precision, the scattered electrons behave as if they are being emitted coherently in space by the individual atoms. And just as the two slits in the photon experiment act as coherent sources of photons, and create an interference pattern, so the nickel atoms act as coherent sources of electrons, and likewise create an interference pattern that can be detected by placing some sort of detector (a photographic film will do fine) adjacent to the nickel crystal, to record the scattered electrons.

The fact that these experimenters saw an interference pattern means that the electrons displayed wavelike properties. You can't make interference patterns with particles. The rule for electrons, as with photons, is that their wavelength depends on their energy, or equivalently on their momentum; the higher the momentum, the shorter the wavelength. Because the electrons all had the same momentum, they created an interference pattern from which a wavelength could be deduced, and the wavelength was exactly what quantum mechanics demanded. (It's important that all the electrons have the same wavelength; if not, interference patterns from electrons with different wavelengths would overlap in space, blurring together the bright and dark areas into an uninformative smudge.)

Since 1927, the wavelike aspect of objects traditionally thought of as particles has been demonstrated many times, by many people. Atoms, like electrons, can be scattered, and can create interference patterns. Just recently, a version of the two-slit experiment was done with atoms instead of photons, and the appropriate interference pattern emerged.

Such results are commonly thought to uphold the idea of "wave-particle duality," according to which fundamental objects are neither waves nor particles, but sometimes one thing or the other, or perhaps always a little of both. This has been held to be one of the fundamental puzzles of quantum mechanics, and although, historically, it may be just that, there's little to be gained by focusing on it as if it were somehow the epitome of quantum mechanical weirdness.

What we're learning is that the phenomena you see in any experimental arrangement depend not just on the things you are measuring, but on the things you set your apparatus up to measure. If you send an electron through a Stern-Gerlach magnet, it acts like a particle, coming out in one beam or the other. But if you scatter electrons coherently from a crystal, they behave like waves, creating an interference pattern. To then agonize over whether electrons are really waves or particles is fruitless. The key is to take note of what you measure, and not try to infer from that what is going on behind the scenes.

Or think again of what happens when you measure an electron's spin with a Stern-Gerlach magnet that can be set either vertically or horizontally. We've seen that you will get an up-down or a left-right measurement, and that it's useless to go beyond that by pretending you can figure whether the electron's spin, prior to measurement, was "really" of an up-down or a left-right character. In fact, it's neither of these, it's indeterminate; it's the measurement itself that defines the thing you measure. There's no purpose in trying to formulate a principle of "up-down/left-right duality" for electron spin, because you're trying to pin something down which is quite indefinite. But in terms of fundamental quantum mechanics precepts, "wave-particle duality" is exactly the same issue. You can measure one kind of property or you can measure the other, and having seen how that works you can see how pointless it is to try to figure what an electron, or a photon, or an atom, is "really" like. Wave-particle duality is by all means an example of how quantum mechanics works, but it is just one example of the pervasive and sometimes frustrating principle that what you get is what you measure. Asking whether objects are really particles or really waves is simply not a meaningful inquiry.

Which way did the photon go?

Now that we are thoroughly acquainted with the idea of wave-functions, let us look again at the riddle of the two-slit experiment. The problem was this: even when single photons are traveling alone through the apparatus, they manage to hit the screen on the other side of the two slits with the correct distribution to create, over the course of time, the familiar stripes of an interference pattern. Each photon manages to generate a piece of the overall interference pattern all by itself.

Common sense inclines us to think that the photon, being at least some sort of a particle, must go through one slit or the other, except that the appearance of any kind of interference pattern seems to require that something go through both slits. How are we to understand this?

To begin, what is the significance of doing the experiment at very low intensity, so that photons travel through one by one, rather than at high intensity, so that photons are streaming through the two slits on each other's heels?

As a matter of fact, none at all.

Each single photon that goes through the apparatus is described by a wavefunction that is interpreted as the probability for that photon to hit the screen in one place or another. Evidently, the same wavefunction describes every photon: they are identical, and their pattern of probability is the same. So when we run the experiment, photons travel through it to hit the screen, and the wavefunction ensures that their pattern of impacts builds up to create the interference stripes.

But this is equally true whether a million photons pass through the two-slit experiment in a second, all jostling through together, or if they pass through the apparatus in (let us be fanciful) a million years, one reaching the screen every year. Either way, the same pattern is left at the end of the experiment, whether it took a second or a million years to build up. It may seem to us that the second case, where the photons are clearly isolated from each other, is more perplexing, but the physics is the same in either case: a million separate photons, each independently described by the same wavefunction, create an interference pattern.

The rules of quantum mechanics are such that the precise path of no individual photon can be predicted; we can only say, having calculated the appropriate wavefunction, that each photon has a certain probability of arriving at any given spot on the screen. When a million photons have struck the screen according to this distribution of probabilities, the stripes appear.

We may like to think that when a million photons are passing through the apparatus all at once, they are all somehow conspiring (one photon whispers to another, if you go there, and I tell that one to go over there, and I move a little over here, then everything will come out right. . . .) to produce the required result. But when the photons are coming through at the rate of one a year, there is no opportunity for this kind of conspiracy, and we are puzzled, therefore, how the stripes arise.

But in fact, there is no photon conspiracy in either case. In neither case are photons interacting with each other as they pass through the slits; even when there are millions of photons in the system at the same time, the laws of physics are such that the photons do in bulk exactly what they would do in isolation. There is no interaction of any kind between the photons in the two-slit experiment. It is wrong and misleading to think that the photons are any less isolated when a million of them are in the apparatus together than when they pass by at a year's interval from each other. Photons do not interact in this experiment; they are always alone. Whether photons arrive at the photographic film once a year or once every millionth of a second, it is still the case that each photon creates a localized bright spot, and that the sum of all those separate bright spots is what constitutes the striped interference pattern.

The virtue of thinking about an experiment when photons are manifestly alone as they travel through the apparatus is that it forces us to think more clearly about the way this interference pattern arises. And we conclude, to our dismay perhaps, that it is just as hard to understand what is happening in the apparently simpler case, when lots of photons are in the experiment at the same time.

Each individual photon travels through the apparatus in the same way, regardless of whether there are lots of other photons around or not, and it does so, moreover, in a manner that depends crucially on the existence of both slits. No matter how we word it, we have to conclude that each photon is somehow "aware" of the presence of the two slits rather than one. If we want to insist that any photon must actually pass through one slit or the other, we are still obliged to acknowledge that the photon, as it passes through one slit, behaves as if it were somehow cognizant of the fact that it could have passed through the other instead.

•

If this is beginning to sound mystifying, it is because we are near the heart of what makes quantum mechanics difficult to grasp. Is there a way we can modify the two-slit experiment so as to find out which way the photon went? This turns out not to be an easy task.

First, if you simply cover one of the slits with a piece of tape, you can be sure any photon has to go through the other slit. But of course the interference pattern disappears: there is now only one possible photon path, and interference requires two. This is not very instructive.

What if you keep both slits open but install, just behind one of the slits, a photon detector? Suppose the detector is of a sort that catches a photon and makes a signal, so that the photon is then lost. You will find that the detector registers a photon half the time, indicating that, on average, a photon goes through this slit 50 percent of the time the experiment is tried. The other 50 percent of the time, a photon presumably goes through the other slit.

But, as before, the interference pattern vanishes. This is a bit more illuminating. If the detector registers, then a photon has gone through that slit and been caught; there was only one photon in the apparatus to begin with, so in this case no light reaches the screen. If, on the other hand, the detector fails to register, the photon must have gone through the other slit; but in this case what you have is a single photon going through a single slit, which cannot produce an interference pattern.

But we have learned something new here. If you alter the experiment in such a way as to require information about which way the photon went, then the interference pattern goes away.

Ideally, you would like to set up an experiment in which you have a device that tells you which way the photon went, but which then lets the photon travel on its way undisturbed. As stated, this is impossible to achieve: any device that detects a photon, no matter how subtly, must affect it in some way, so that the experiment is fundamentally changed, and the interference pattern (which requires parallel and coherent existence of the two photon paths) goes away. But in this impossibility lies an important point. Anything you do to force an answer to the question "did a photon go this way or that?" causes the interference pattern to vanish.

Ultimately, this is another example of the uncertainty principle. Either you can observe an interference pattern or you can find out which way the photon went, but you can't do both. Seeing the interference pattern and detecting a photon are both measurements, so here again are two measurements that cannot both be performed in the same experiment—you have to choose one or the other. You can have one piece of information, or you can have the other, but you can't have both.

(This is also another example of wave-particle duality. Detecting which slit the photon went through is a particle measurement; recording the interference pattern is a wave measurement. You can do one or the other, but not both. But calling it an example of wave-particle duality rather than of the uncertainty principle obscures the generality of this kind of behavior.)

Here's a better way of thinking about all this. In the standard two-slit experiment, asking "which slit did the photon go through?" is a meaningless question. You might like to think, because it makes you feel comfortable, that any particular photon has to go through one slit or the other, or you might persuade yourself, on the other hand, that you feel comfortable with the notion of a photon wavefunction sufficiently spread out that it can sample both slits simultaneously. It really doesn't matter: you don't have the means to provide an answer to any such question, and the issue is moot. But you see the interference pattern.

Now, however, when you place a detector next to one of the slits, you are forcing a choice. The detector either sees a photon or it does not. It certainly can't record a fraction of a photon.

Asking "did a photon pass this way or not?" produces, of necessity, a yes or no answer. By demanding this additional information you are requiring the photon to behave in such a way that it must be seen to go through one slit or the other. In those circumstances, no interference on the far side of the slits can possibly happen, and the interference pattern does not appear.

Or here's another way of thinking about it. The idea of a thing called a "photon" has useful meaning only in the context of a measurement that can say whether a photon is there or not. The little detectors that we arrayed on the screen to record the interference pattern register electronic changes one photon at a time. Therefore, when you use quantum mechanical formulations to arrive at a description of the pattern that appears on the screen, you are obliged to couch that description in terms of photons. Similarly, when a photon detector is placed next to one of the slits, and you use quantum mechanics to describe how the detector will respond, you are obliged to formulate your answer in terms of the passage of a photon—or not— through that slit.

But when you are contemplating what is happening in parts of the apparatus where there is no device to record photons, the very concept of "photon" makes no sense. So in the unadorned experiment, where no detector is placed next to either slit, it is misleading to say that you cannot tell whether a photon went this way or that, because that phrasing implies that real photons are flitting about, except that you do not have the means to tell where they are. The more accurate thing to say is that if no device is present that can detect photons, asking whether a photon is there or not is meaningless.

In fact we already said, a few paragraphs above, that this is a meaningless question. There's a subtle distinction, though: up above, we were implying that asking whether the photon went this way or that is a reasonable question, but one that must be set aside unless you actually go to the trouble of obtaining an answer. Now we are saying that it is not even a reasonable question.

We are sneaking up on the idea that a photon is not real until you measure it, or perhaps that it is the act of making a suitable measurement that causes a photon to be present or absent. This is the same issue that arose early on when we thought about measuring the spin of an electron. When a measurement takes place, the spin must be either "up" or "down." But before the measurement takes place, or in the absence of any measurement at all, is it better to say that the spin has some indeterminate value, or that "spin" is meaningless until you declare the means by which you intend to define it?

The idea that "spin" and "photon" become meaningful words only when you make a measurement of them is part of a constellation of ideas that comes under the heading "collapse of the wavefunction." The notion here is that an unmeasured wavefunction—such as applies to a photon, for example, in the space between the light source and the screen in a two-slit experiment—can encompass a range of possibilities that have no classical meaning: a photon being partly here, partly there; a spin being half-up, half-down. It is only when you make a measurement (did a photon go through this slit? is the electron up or down?) that the wavefunction is obliged, so to speak, to yield a definite answer. What was initially a photon wavefunction spread out over the screen becomes an actual detected photon in one of the detectors arrayed on the screen, and an absence of any photon in all the rest. What was initially a half-up, half-down electron becomes simply an up electron. After any such measurement, the wavefunction becomes less expansive or capacious than it was. Hence the name "collapse" (sometimes "reduction") of the wavefunction.

Whether this collapse is a physical or a psychological event is not easy to say.

No, but really, what really happened?

If we want to know which way the photon went, we can't see the interference pattern. If we want to see an interference pattern, we can't ask to know which way the photon went. This will be a dismaying conclusion to those who believe that we are entitled to find out whatever we want about any physical system, but it does not yet seem paradoxical. A variation on this sort of experiment does seem to lead into muddier waters, however.

An ordinary piece of glass, held at a suitable angle, can both reflect and transmit light; think of seeing, in your car window as you drive along an urban street at night, reflections of your backseat passengers' faces looming over the street scene outside. If light is a stream of photons then we must conclude, yet again, that each photon must be either transmitted through or reflected by such a piece of glass, with the appropriate probability.

By suitable elaboration of this idea, the physicist can make a beam splitter, a device that sends photons one way or another with equal, fifty-fifty probability. If light from a laser is thus divided, the two beams are physically distinct but nonetheless coherent, in that their wave motion beats up and down in unison. If the two beams are guided, with mirrors, to recombine on a screen, then an interference pattern will appear. Though the technology is a little different, this is really no more than a version of the two-slit experiment. As before, the perplexing point is that even if photons are sent through one at a time, so that the beam splitter, one would think, must send the photon by one route or the other to the screen, an interference pattern appears nonetheless. The single photon somehow travels both routes, just as it seemed to go through both slits.

And, predictably, if you put a detector in one path or the other, the interference pattern will disappear.

This new arrangement allows for a somewhat different kind of experiment which (to some people's way of thinking anyway) shows up even more puzzlingly the strange nature of the photon. The two routes along which the photons travel can be, within reason, as long as you like. This means you can now install a photon detector on one of the pathways at some considerable distance from the beam splitter—distant enough, at any rate, that you can wait until the photon has incontrovertibly passed through the beam splitter before deciding whether you are actually going to switch the detector on or not. This allows for what has become known as a "delayed choice" experiment. With the detector off, we have the standard means of creating an interference pattern, but with the detector on, we are asking which way the photon went, and therefore lose the interference pattern. The new ingredient is that our choice of which experiment to perform is not made until the photon is en route, having passed through the beam splitter.

We might think that in an interference experiment, the photon mysteriously travels both routes at once, but that in the "which way?" experiment, the photon is forced to take one path or the other. But if we delay our choice of which experiment to perform until after the photon has gone through the beam splitter, how does it know what to do?

Let us try to be more precise about the question we are exploring here. The argument just stated hints that the photon somehow knows, when it gets to the beam splitter, whether it is destined for an interference pattern or a "which way" detection, and behaves accordingly. When we add the notion of a delayed choice, this apparent premonitory aspect is exaggerated to a bizarre degree.

Consider: suppose that the pathways are long enough to allow someone to toss a coin in order to decide quite randomly whether the detector will be switched on or not. Now, the physicist sets the experiment in motion and, at the time the photon reaches the beam splitter, gives no clue (hands in pockets, whistling tunelessly, staring at the ceiling) what is going to happen next. Then as soon as the photon has passed through the glass plate—having committed itself, we might suppose, to either traveling for sure down one path or the other, or else dividing itself up equally between both—our experimenter rushes to the door, drags in a small boy standing outside, gets him to toss a penny, and, if the penny comes up heads, switches on the detector.

With the detector switched on, we are dealing with a "which way" experiment, and the photon is required to be in one pathway or the other. If, on the other hand, the penny had come up tails, the detector would have been left inactive, the photon would have explored both pathways, and we would be dealing with an interference experiment. In short, we seem to have devised an experimental arrangement in which the photon, when it passes through the glass plate, is required to choose one kind of behavior or the other, before anyone, not even the experimenter or the small boy with the penny, knows which kind of behavior will subsequently be required.

And yet, when an experiment of this sort is performed, the photon is invariably found to have made the right choice. If the penny came up tails and the detector was not activated, an interference pattern appears; if it came up heads, and the detector is switched on, interference goes away. Somehow the photon gets it right. Even if you believe in the possibility of clairvoyance, it is galling to think that a mere whisper of electromagnetic energy can predict the toss of a coin 100 percent of the time, when even the most die-hard of ESP enthusiasts gloats over a 52 percent success rate.

But if we find it absurd to start thinking that photons can

predict the future, we had better think again about how we interpret this experiment.

The answer is to go back to the statement with which we began this section: If we want to know which way the photon went, we can't see the interference pattern. If we want to see an interference pattern, we can't ask to know which way the photon went. This is the minimal interpretation we can put upon any experimental arrangement of this sort, and it still holds true, even in the so-called "delayed choice" experiment. No matter what sort of elaborate preparations were made, no matter how randomly the coin tossing was done, the fact remains that in the end, the detector was either switched on or it was not. And an interference pattern was not observed, or it was, respectively. This simple logic includes all the unarguable experimental data derived from the experiment, and it makes valid connections between them. If we stick to this simple logic alone, nothing remarkable emerges from the delayed choice experiment that was not already seen in the two-slit experiment.

The problems arise only when we surrender to the urge to start imagining, once we know what the outcome of an experiment is, that we can deduce accurately what must have happened along the way. That is, if we see an interference pattern, we think that the photon must have divided itself into separate pieces that traveled each pathway, whereas if we know that the photon was detected along one pathway, we find it impossible to resist the thought that it must have actually gone down that pathway and not the other. But then when we do a delayed choice experiment, we find that insisting on one or the other of these two mutually exclusive interpretations leads us into trouble, because it appears that the photon must know in advance what is going to happen, so that it can choose the appropriate behavior to follow. But, and this is the fundamental point, those two possible behaviors are not actual behaviors that we know the photon must in fact have followed, but inferred, deduced, or (more accurately) speculated behaviors, which we have attributed to the photon in order to make some sense, so we think, of the experimental results.

It's a legitimate, indeed, essential facet of science that one makes deductions or hypotheses from one set of experimental data, and devises more elaborate tests to see if those hypotheses are reasonable. The lesson of the delayed choice experiment is that any hypothesis that forces the photon to adopt one or the other of two distinct and contradictory behaviors is, in fact, not reasonable.

And we should have known this in advance. What was the first thing we learned about quantum measurements, from the Stern-Gerlach experiment? That the spin of an electron is strictly an undetermined quantity until an experiment yields a measurable value for it. And, moreover, that any attempt to say, having found an electron to be in an "up" orientation, that it must have been that way all along, is demonstrably incorrect, because any such deduction is incompatible with the idea that a spin measurement along a different direction would have yielded, for example, a left-right distinction rather than an up-down one. To repeat, it is going beyond the bounds of quantum mechanics to think that you can deduce, from any measurement, what the prior state of the measured system "really was." Rather, the system is indeterminate until the measurement is made.

And the delayed choice experiment brings that point forcefully into the open. Seeing an interference pattern, or detecting a photon in one path or the other, is at bottom a simple measurement made on a complicated system. And any attempt to think that we know, once the measurement is done, what really went on inside the system (whether the photon went one way or the other, or both ways at once) is precisely an attempt to pin down the prior state of a system after a measurement has been made. It causes trouble, and we must resist the temptation to do it.

The minimal interpretation of quantum mechanics embodied

in the first two sentences of this section is like a wise but stern parental injunction against certain kinds of teenage behavior: limiting, to be sure, but it keeps people out of danger.

How to make money from quantum mechanics

The Hungarian financier and speculator George Soros made hundreds of millions of dollars in the late 1970s and early 1980s by means of a financial investment device called the Quantum Fund. There was more to the name than its attractively modernistic sound. Soros had written books expounding a personal and philosophical view of history, and his activities in the world's stock markets were supposedly governed by the principle that "participants are not detached observers. Their thinking affects the situation to which it refers."

This philosophy has something to do with the quantum mechanical idea that measurement affects the thing measured, so that Soros was supposedly taking into account, in his investment strategies, the fact that his actions had some influence on the very markets he was investing in, rather than adhering to a more classical philosophy which would hold that the market moved to its own tune and investors were mere bystanders, hoping to derive financial gain from market movements over which they had no control.

There arc two problems with this. First, investors are well aware that their actions influence the market; what indeed is the stock market but the sum total of the activities of all its participants? The fact that investors have from time to time sought to corner the market in one commodity or another betrays a realization that the market can in principle be controlled.

More pertinently for the subject of this book, the notion that measurement affects the thing measured is hardly unique to quantum mechanics. Any measurement of a physical system necessarily involves an interaction between that system and a measuring device, and any interaction means a trading of influences back and forth. Think of using a thermometer to take the temperature of water. If you place a thermometer at room temperature into a bucket of warm water, the mercury in the thermometer heats up, and its expansion inside the thermometer tells you how hot the water is. But if the mercury rises because it is heated by the water, then inescapably that amount of heat energy was taken from the water, which perforce must become cooler.

Of course, a small thermometer has an almost immeasurably small effect on a large bucket of water. But think of using an ordinary household or doctor's thermometer to measure the temperature of a teaspoonful of hot water. The very act of temperature measurement means placing the thermometer in the water and leaving it there until the temperature of both water and mercury are the same. But if the volume of water is not so very different from the volume of mercury in the thermometer, then the temperature of the water may fall by about as much as the temperature of the mercury rises, so that what you eventually measure is some sort of average of the two—not the temperature that the water was before you put the thermometer in it. Clearly, the process of making a measurement has altered the thing you wanted to measure.

But here's the difference between classical and quantum physics. If you know exactly how much water was in your teaspoon and how much mercury was in your thermometer, and if you go to the library and look up a physical quantity called the heat capacity of the two liquids, which measures how much heat energy you have to put into a quantity of water or mercury to raise the temperature by one degree, then you can work out, from your measurement, exactly what the temperature of the water must have been before you put the thermometer in it. With the right additional knowledge, in other words, you can compensate for the clumsiness of the measuring process and calculate with considerable confidence the number you were after.

A different and more practical strategy would be to acquire a smaller thermometer which would not disturb so greatly the temperature of the water; perhaps you can obtain some sort of electronic device (rather than an old-fashioned mercury thermometer) that necessitates placing only a tiny sliver of an electrode into the water. With sufficient ingenuity, you can come up with a thermometer that will measure the temperature of a teaspoonful of water without any significant disturbance to the system, and so get as accurate a result as you desire.

In quantum mechanics, neither of these strategies will get you anywhere. If you are measuring electron spins in a Stern-Gerlach magnet, the process of measurement forces what was an indeterminate quantity to take on a specific value. When the measurement is made, an electron's spin must be either "up" or "down," and no meaning can be attached to the idea of trying to infer what particular state it was in before you made the measurement.

Or if you try to measure the position of an electron, as in Heisenberg's experiment, by bouncing a photon off it and measuring the recoil, there is an unavoidable element of chance in the encounter, so that even after you have measured the energy and direction of the recoiling photon, you can say only that the electron had a 90 percent probability, for example, of having been in a certain volume of space when the photon interacted with it. Changing the energy of the photon can give you better information about either the position or the speed of the electron, but there is always a compromise.

It is this element of probability that makes quantum measurement different from classical measurement. When you specify that an electron has a fifty-fifty chance of being spin-up or spin-down on emerging from a Stern-Gerlach magnet, you have said all you can possibly say about that particular measurement. And after you have measured a couple of electron spins, and found that one was up and one was down, you can still infer nothing about any hypothetical difference between those two electrons before they were measured. The fact that they came out in two different states implies no difference at all in their states before measurement. There was no additional information you could acquire that would have told you what the spin measurement of any particular electron was going to be—apart, of course, from actually doing the measurement itself, which then renders further measurement superfluous. Probability cannot be expunged from quantum measurements.

In fact, it is misleading to say that "measurement affects the thing measured" because that can seem to imply that a quantum object was in some definite but unknown state, but was then disturbed by an act of measurement and is now in some other state. Rather, measurement gives definition to quantities that were previously indefinite; there is no meaning that can be given to a quantity until it is measured.

The importance of being rigorous

We seem to have reached the point where we can enunciate a principle: nothing is real until you measure it. Or, as Niels Bohr's protégé John Wheeler liked to put it, "No elementary phenomenon is a real phenomenon until it is a measured phenomenon."

The idea that you can deal only in measured or observed quantities may not seem like a strong injunction. Isn't that what scientists do all the time? In fact, it's not. Scientists have almost always assumed that what they are measuring is part of some deep and invariable reality, that there exists an objective world that we can, by degrees, apprehend. Quantum mechanics says no to this: if we try to impute some sort of physical reality to the wavefunction, for example by declaring that a photon really does travel down both parts of an interference experiment, then we have difficulties understanding the delayed choice experiment. Bohr's injunction, the central pillar of what has become known as the Copenhagen interpretation of quantum mechanics, is a good deal stricter than it may seem. It positively forbids us from imagining that we know what the photon did as it went from one end of the interference experiment to the other. But imagining that they know, from observable data, what is going on "inside" a physical system is precisely what scientists—and physicists in particular—have generally assumed they could do without inhibition or penalty.

Seen this way, the Copenhagen interpretation of quantum mechanics is not so much a philosophy as an act of intellectual self-discipline. It does not, you may well say, make the two-slit experiment any easier to understand; it tells us we should not hope to understand it in the way we would like to. It resolves certain difficulties only by declaring them out of bounds. There's nothing to say that you can't do more complicated experiments and try to extract more information, only that you cannot both do that and at the same time expect your original conclusions to remain unchanged. Different experiments, different results.

The Copenhagen interpretation is, fundamentally, the uncertainty principle writ large. In its simplest form, the uncertainty principle puts limits on what we can know: you can't know both the speed and position of an electron; you can't measure its spin in both an up-down and a left-right sense at the same time. More elaborately, you can't ask to see an interference pattern and also know which way the photon went. And finally, you can't infer what's "really" going on in one kind of experiment and expect it to be consistent with what's "really" going on in a modified, and therefore different, version of that experiment. That's the Copenhagen interpretation, more or less.

If, with Bohr, we acknowledge that inferences from data are

not always reliable and consistent, but that we will be all right as long as we stick with the data, the whole data, and nothing but the data, are our interpretational difficulties at an end?

No. There's a big problem we haven't yet addressed, and which in fact the Copenhagen interpretation doesn't begin to address. What exactly do we mean by a measurement?

The chronic poor health of Schrödinger's cat

We have established that physical properties are not, as in the classical world, intrinsic and unchangeable characteristics of the things we are measuring but instead arise, in the quantum world, as a result of the act of measurement, and cannot be ascribed any useful or consistent meaning before a measurement is made. This neatly divides the world into two types of physical objects: things we measure (or might in principle measure, should we so wish) and things we do the measuring with. These two kinds of things seem to be fundamentally different. Objects to be measured live in uncertain, indefinite, fuzzy (whatever inadequate word you want to use) states until they are measured, at which point they are obliged to assume one particular characteristic or property or value out of the range of properties they might have had. Objects that do the measuring, on the other hand, exist always in definite states: a photon detector has either detected a photon or it has not, and there is no ambiguity which state it is in. Using a Stern-Gerlach device, we would in practice have some sort of detector along both the "up" and "down" trajectories of the magnet, and we could be sure, when an electron passes through, that it will emerge and be registered in either one detector or the other. There is no intermediate state, no indefiniteness or ambiguity.

We have come across this distinction before. Objects to be measured are quantum objects, with all the attendant indefiniteness and uncertainty and potentiality that we are slowly learning to live with. Objects that measure, by contrast, are unambiguous and actual, always definitely one thing or another. They are, in fact, classical devices, with fixed and incontrovertible characteristics.

This is the fatal inconsistency of the Bohr or Copenhagen interpretation of quantum mechanics. It asserts that quantum objects can be ascribed definition only when a measurement is made, but to make a measurement you need some sort of classical device that gives you an unambiguous result. And yet we also want to believe that quantum mechanics is the fundamental theory of physics, so that, in principle, the inner workings of all things in the world ought to be explicable in quantum mechanical terms. In a modern laboratory, for example, detectors will often be electronic devices hooked up to computers, all depending on the motion and interaction of electrons for their proper functioning. So if we are using electronic devices to make measurements of other electrons, are we to think that the electrons in our detectors and computers are somehow immune from the uncertainty that all too plainly afflicts the electron about to enter the Stern-Gerlach magnet?

In the plainest terms, the Copenhagen interpretation of quantum mechanics relies on a notion called measurement, which is quite beyond the abilities of quantum mechanics itself to explain. Measurement has to be added in by blunt assertion. Bohr was well aware of this problem, but made the pragmatic point that physicists, when it comes down to it, know how to measure things. Laboratories are full of stolid devices that provide reliable and unambiguous results. Physicists had been measuring things for years before quantum mechanics came along, and continued to measure things afterwards, with no sudden loss of confidence.

Pragmatically, physicists understand measurement just as lawyers understand pornography and philistines understand art: they can't define it, but they know what it is. Niels Bohr relied explicitly on this long-standing ability. He never defined, in quantum mechanical terms, at what point a measurement was made and at what point, therefore, what was previously indefinite became definite. He merely observed that, in a practical sense, knowing when a measurement has been made never presented a problem. He trusted his fellow physicists to know what they were doing.

Nevertheless, he and many others were well aware that this reliance on a process that could not be defined represented a philosophical, if not a practical, flaw in his way of thinking. And there was always the danger, as experimenters became technically more ingenious, that they might find places where it became necessary to define with some precision the boundary between the measured and the measuring worlds—the boundary, in other words, between quantum and classical physics.

In an effort to illuminate more clearly this conundrum, or perhaps to show in some specific way why the Copenhagen interpretation must be wrong, a number of physicists used the measurement problem to try to devise "paradoxes" (although, in the end, these demonstrations were undoubtedly strange and thought-provoking but never exactly paradoxical). The most notable of these illustrations is the one thought up by Erwin Schrödinger, an Austrian physicist who largely invented the wavefunction, along with an equation that determines its behavior. Like Planck, Schrödinger remained uncomfortable with what he had brought into the world. To present his "paradox," Schrödinger enlisted the assistance of a cat.

Schrödinger's cat puzzle can be formulated in numerous ways. To keep things as familiar as possible, let us imagine a version based on something we are by now comfortable with—the Stern-Gerlach measurement of an electron's spin. Imagine placing into a closed box a cat, and a device that will automatically perform a spin measurement on an electron. The Stern-Gerlach device is set up in such a way that if the electron comes out in the up beam (which will be registered by a suitable detector in the beam path) then the cover is lifted from a bowl of cat food, but if the electron comes out in the down beam, a separate detector triggers the breaking of a capsule of poison gas, and the cat instantly (and painlessly, please) dies. (See Figure 5.)

The experiment can be controlled by a timer, so that we can put the cat and the device into the box and arrange for the spin measurement to be made after half an hour. Then imagine we wait another half hour before looking into the box to discover the cat's fate. Clearly, the cat has a fifty-fifty chance of being dead or alive when we open the box, depending on whether the electron happened to come out "up" or "down."

The facts are simple. The question is what we make of them. Our first inclination is to say that at the half-hour mark, a spin measurement was made, and the cat either survived or was killed. After that, we know for sure the cat is either alive or dead, but we don't know which until someone looks inside the box.

But to say this implies, now that we are thinking quantum mechanically, that the spin detector is a classical measuring device, definitely recording the electron's emergence in one beam or the other, and triggering the appropriate consequences. Before the measurement, the electron was in an indefinite "half-up, half-down" state, and afterwards it was either "up" or "down." But until now, at least, we have more or less taken for granted the idea that a measurement takes place when some piece of information becomes available to the watching physicist. Can we be sure that a Stern-Gerlach device in isolation, unobserved, genuinely constitutes a measuring device?

Here's another possibility: since the Stern-Gerlach device itself is presumably functioning according to the laws of physics, and in particular according to the strictures of quantum mechanics, what is there to stop us from trying to describe it in quantum mechanical terms? We could say that the detector has two quantum mechanical states, namely UP and DOWN. The

FIGURE 5

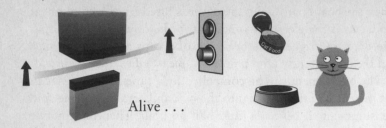

Alive . . .

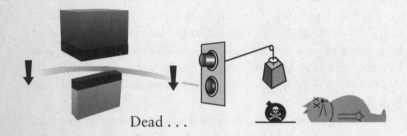

Dead . . .

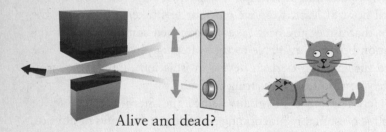

Alive and dead?

If the electron is measured to be "up," the cat gets fed; if it's measured to be "down," the cat expires. But does a "left" or "half-up, half-down" electron create a "half-alive, half-dead" cat?

way it works is that an up electron triggers an UP response in the detector, and a down electron triggers a DOWN response.

But then, must not a half-up, half-down electron trigger a half-UP, half-DOWN response in the detector? Remember, we are now conducting an exercise in which we are deliberately describing the detector as a quantum mechanical system, not as a classical detection device in the Copenhagen sense. It's not that we are obliged to do this; just that the Copenhagen interpretation does not seem to specifically forbid us from doing so, and being good and adventurous physicists, we should always explore the consequences of any argument that is not explicitly ruled out.

And so, if we think that the detector is a quantum system, there must be some other, larger device to measure *its* state. Such a device might be you, when you look inside the box. An observed detector, we are fairly sure, cannot be in an uncertain half-UP, half-DOWN state, so that at the very least we would have to conclude that when the box is opened and the Stern-Gerlach device itself is observed, it must resolve itself (through the agency of observation) into either an UP or a DOWN state.

But does this then mean that after the electron was made to pass through the magnet (and we know for sure this happened at the half-hour mark) but before anyone looked inside the box, the cat itself was in an uncertain "half-dead, half-alive" state? After all, we have tied the cat's state to the state of the electron-spin measuring device, so if a definite measurement is not made until the box is opened, must not everything inside it, cat included, be in an uncertain, mixed, half-one-thing, half-the-other state until that time?

It's a bit of stagecraft, you may realize, that Schrödinger put a cat into this scenario. Imagining a full-scale Stern-Gerlach electron-spin detector in an undecided half-UP, half-DOWN state is really no more or less difficult than imagining a cat in an equally uncertain half-dead, half-alive state. Until this point, we were getting somewhat used to the idea that an electron could be in a half-up, half-down state, without ever being able to get a clear idea what that really meant, for the elementary reason that any electron we ever see must always be in one state or another. The uncertain half-up, half-down state can by definition apply only to electrons whose spin we have not yet measured.

Yet perhaps we can accept that elementary particles such as electrons, which we can't see anyway in the way we can see a billiard ball or a glove, can behave in ways that don't really make sense to us, simply because they are objects distant from our apprehension in the first place. Cats are a different matter. We see cats all the time, and we know that cats (among other things) are always either alive or dead. We don't know what it means to say that a cat is "half-dead, half-alive" any more than we really know what it means to say that an electron is "half-up, half-down" or that an electron-spin detector is "half-UP, half-DOWN."

So what exactly are we to learn from the tale of Schrödinger's cat? One way to look at it is as an illumination of the fact that, indeed, uncertain quantum mechanical states are hard, perhaps impossible, to visualize. We can still take comfort in the fact that the cat, once the box is opened, must be definitely either dead or alive; we can never see this purported half-dead, half-alive cat, and so, if we follow the Copenhagen interpretation with absolute rigor, we should simply decide not to worry about what it means, since it refers to a state we can never detect.

But taking this point of view only emphasizes the fact that the Copenhagen interpretation offers no guidance at all on how a measurement is actually accomplished. It may be that Schrödinger's cat does not present us with any genuine paradoxes, in that the final outcomes of measurements and observations are never in dispute, but if we have to accommodate the issues that arise by deciding that the act of measurement or observation occurs whenever we want it to occur (at the half-hour mark, when the electron passes through the Stern-Gerlach magnet, or at the hour mark, when we open the box, it really makes no difference) then we seem to be reducing a fundamental ingredient of physics—measurement—to a more or less psychological phenomenon. Can this be possible?

Psychophysics, *qu'est-ce que c'est?*

Thinking about the way an electron in an uncertain state becomes, when measured, either an "up" or a "down" electron led us to the idea of the collapse or reduction of the wavefunction: the "half-up, half-down" wavefunction before measurement reduces to simply an "up" or a "down" wavefunction after measurement.

What the story of Schrödinger's cat illustrates is that we don't know at which point along a chain of measurement and detection the wavefunction collapses—at what point, in other words, properties are transformed from fuzzy quantum indeterminacy into reliable classical knowledge. The Copenhagen interpretation gives us no help, and Bohr tells us only that as a practical matter, it isn't an important question. Always, in real experiments, it's unambiguous that a measurement becomes definite at some point, even if we can't say exactly where, and our ignorance of precisely how and where a measurement gets made doesn't seem to have any real consequences. Nevertheless, this state of affairs is metaphysically discomfiting, and it's not beyond the bounds of reason to think there might be consequences. For measurement to work the way Bohr needs it to work, it seems as if measurement itself is in some sense (undefined, so far) a non-quantum-mechanical process. Could this really be the case?

•

Broadly speaking, there are two ways to deal with Schrö-dinger's cat. The first is to say that we know, in a laboratory, that Stern-Gerlach magnets work unambiguously as measuring devices, reliably generating UP or DOWN results, and never getting suspended in some incomprehensible and ill-defined half-UP, half-DOWN state. There's no reason to think that such a device would physically work in a different way just because its operation happens to be concealed from direct inspection, so we assert that, in the box with the cat, the measurement incontrovertibly takes place at the half-hour mark, at which time we know the electron is sent through the Stern-Gerlach magnet. A measurement takes place, the cat is either dead or alive, and it's just a matter of time before someone opens the box and finds out which. This is no more puzzling than tossing a coin and having it roll under your upright piano, where you can't see it, and having to wait for your well-muscled brother-in-law to arrive and lift the piano up for you to see whether it's heads or tails.

The problem with this point of view is that it leaves entirely unexplained what constitutes an effective measuring device. Does it just have to be sufficiently large, containing enough atoms that the individual quantum behavior of those atoms somehow doesn't matter? But how many atoms? And anyway, why should quantum behavior go away just because a lot of atoms are connected together to make a single device? Or do we want to imagine that quantum mechanics is actually a correct theory only for individual objects and particles, and that some specific quantitative modification is needed to explain larger systems?

If we just insist on declaring that measuring devices do in fact work, and never mind how, then we can forget about Schrödinger's cat. But this is a distressingly arbitrary and mysterious declaration to make: it amounts to saying that quantum mechanics is an excellent, detailed, and complete theory, but that we need a tooth fairy to make sense of it.

The second strategy is to say that measurements become real only when someone takes note of them. This seems a somewhat whimsical attitude to apply to a theory of physics, and it leads into all kinds of philosophical swamplands. Eugene Wigner came up with one argument, involving a fictitious friend of his. Suppose that Schrödinger's cat is in its box, and that the requisite time has gone by, so that the cat has a fifty-fifty chance of being dead or alive. Now suppose that Wigner's friend opens the box and discovers the cat's state—it's either dead or alive— but that he does not yet tell Wigner what the result is. Does that mean that the cat's wavefunction has collapsed from indefiniteness to definiteness as far as Wigner's friend is concerned, but that as far as Wigner himself is concerned, the wavefunction remains uncollapsed until his friend tells him about the cat? Or does the wavefunction collapse for everyone once one person makes measurement or observes a result?

Whichever line of argument we take, we run into trouble. If we say that wavefunctions collapse for different individuals, at the moment they personally acquire knowledge of results, then we seem to be saying that the wavefunction represents not the physical state of a particle or a system but rather our knowledge of that state. Everyone then has a different wavefunction for the world, embodying their particular knowledge of it.

But let us not forget Descartes, who was able to convince himself of his own existence—I think, therefore I am—but not of anyone else's. Why should I be sure of a result or an observation just because some other person (possibly an unreliable person) tells me about it? If I were Descartes, I would believe only in what I am directly aware of, and I would have to conclude that quantum indeterminacy engulfs everything (the results of the Wimbledon men's final, the existence of New Zealand, whether my lawn needs mowing or not while I am away on holiday) that I cannot prove for myself.

And yet, on the other hand, I have been able to function tolerably well in the world by assuming that Wimbledon was won by Pete Sampras in 1995, and that New Zealand does in fact exist. I can't get out of mowing my lawn by telling my neighbor that the grass may seem long in her perception of the world but that according to mine it's neat and tidy.

The puzzle then gets turned around: if the existence of facts depends, for me, on my consciousness, and for you, on yours, how is it that we can agree on so much? How does the fundamental quantum indeterminacy of the world vanish, apparently, when we deal with everyday objects? If the idea that human consciousness transforms muddy quantum fuzziness into solid reliable knowledge was meant to solve this problem, we see that it has not done so at all. We may say that quantum fuzziness disappears, for me, when I observe the world, and that it disappears for you when you observe it, but then we have to acknowledge that it disappears for you and me—and indeed everyone else—in pretty much the same way, which suggests that disappearance of quantum fuzziness has little to do with my observation of the world or your observation or anyone else's.

So perhaps the answer is that anyone's consciousness will do to collapse the wavefunction for everyone: whether Wigner or his friend opened the box and saw the cat first, it doesn't matter. But if we follow this tack we are in effect ascribing some strange and magical quality to the presence of a human observer. And in fact, this is not so different from the first point of view, by which we simply declared that "measuring devices" such as Stern-Gerlach magnets manage to obtain a definite result in some way that seems to go beyond the rules of quantum mechanics alone. Ascribing the same mysterious ability to human consciousness, simply because human consciousness is mysterious anyway, and we might as well throw all the mysteries in one place, doesn't get us very far.

It has been seriously suggested, nevertheless, that human consciousness or perception is somehow the key to the measurement problem; it's not so much the physical act of measurement as the mental act of becoming aware of the result that, ultimately, demarcates the line between quantum uncertainty and specific knowledge. No one has ever suggested a scientific explanation for this strange interaction between wavefunction and brainfunction, but even so a number of reputable physicists (including Wigner himself as well as his hardheaded mathematician friend John von Neumann) have leaned in this direction—more, perhaps, out of desperation than anything else. You may not be able to come up with any objective definition of what constitutes a measurement, but at the very least you can say a measurement is real when someone has observed and noted down the result.

None of these arguments, to sum up, really helps in any specific way to explain what makes a quantum mechanical measurement real. If it's something to do with physics alone, and perfectly ordinary measuring devices are easily able to resolve quantum indeterminacy, then they seem to rely on a kind of physics beyond quantum mechanics. Or if it isn't a matter of physics but of consciousness, then either we are led to an uncomfortable solipsism, in which we each live in a private and enveloping fuzz of quantum indeterminacy but manage to perceive the same solid world nonetheless, or else we simply throw up our hands and conclude that something about consciousness physically interacts with a wavefunction and makes it collapse. But if we don't understand how mere physics can accomplish such a thing, what's the advantage of resorting to psychophysics?

INTERMISSION

A Largely Philosophical Interlude

From the earliest days of quantum mechanics, questions of meaning and interpretation could not be avoided. The unique and infinitely knowable reality of the classical world was shown to be a delusion, but the reality of the quantum world was slippery and elusive. An old world had been lost, but the new world seemed fogbound.

Then again, most physicists didn't lose sleep over any of this. Quantum mechanics, despite its mysteriousness, was both technically demanding and phenomenally successful. Working out the consequences of quantum mechanics, for physical experiments that could be planned and done, consumed the time and interest of physicists much more than grappling with vague concerns about the nature of reality—concerns which might be important but which never seemed to congeal into questions that could be resolved by any practical test.

And so most physicists, during the middle years of this century, knew in the backs of their minds that some things about quantum mechanics were hard to understand, but those mysteries never seemed to cause them any real difficulty in their laboratories. Only a few of the physicists of the (by now) older generation, such as Bohr and Heisenberg and Einstein, along with a handful of younger physicists with philosophical inclinations, such as John Wheeler, ever spent much time worrying over what quantum mechanics meant. Everyone else was too busy using the theory to wonder how it worked.

Nevertheless, Einstein remained unhappy. If the majority of physicists thought that worrying about the philosophical meaning and implications of quantum theory was not quite a proper subject for a real physicist, Einstein at least had earned the right, in his later years, to ponder subjects that distressed him. But through this period, from perhaps the 1930s to the 1960s, thinking about quantum mechanics meant mostly thinking about experiments that were not yet practical, or about ways to tinker with the mathematics of quantum theory so as, perhaps, to render the mysteries a little less mysterious. It was a long time before any of this airy speculation came to earth.

Does the Moon really exist?

The idea that physical quantities don't take on any practical reality until someone measures them hugely offended Einstein's sense of how physics ought to work, and led him on one occasion to ask the physicist Abraham Pais, a little plaintively we may imagine, if he "really believed that the Moon only exists when you look at it."

This is not an easy question to answer.

I have in my mind a theoretical moon. This theoretical moon, a purely mental construct, has certain hypothetical or proposed physical properties: it is a more or less spherical piece of rock, it follows a certain orbit around Earth, its surface has a certain reflectivity, and so on. If someone asks me at any time where the Moon is in the sky, and what phase it presents, I turn to my theoretical moon, make the necessary calculations based on the time of day, the position of the Sun, the latitude and longitude of the observer, et cetera, and I tell my questioner, "If you raise your eyes to this position in the sky, you will find the image of the Moon, and it will have a certain crescent shape whose precise details I can specify, if you wish."

And then my questioner looks upwards, and finds a real Moon in the real sky, just where I said it would be, and with just the shape I predicted.

If, over a period of time, many people test my knowledge of the Moon in this way, they will find I am infallible. My theoretical moon, following its theoretical orbit, always tells me where anyone needs to look in the sky to find the real Moon.

Of course, if it happens to be overcast, I could say that the image of the Moon would be in such and such a place, were it not for the clouds, but since the sky is obscured on this occasion I cannot prove the Moon really is there, and my questioner could not prove that it is not.

Or another more ingenious questioner might tell me that he does not plan to look at the Moon directly, but wishes to observe the shadows cast by the Moon on the ground nearby (it happens to be a clear, bright night). Then I can predict just as reliably as before that the shadows will fall in a certain direction. If this questioner asks me, do I think that a correct prediction of the positions of shadows proves that the Moon is really there, I can only respond that I, as a pure theoretician, have nothing to say on whether the Moon is really there or not, and that if you, my questioner, do not care to look up in the sky for yourself to see if the Moon is there, then the question is moot.

Let us be utterly skeptical. If someone asks me whether I believe the Moon is there even when no one is looking at it, I am obliged to say that the question makes no sense. If you want to verify that the Moon is there, then go ahead and look—but then you are not answering the question. If, on the other hand, you want some proof of the Moon's reality with the stipulation that you are not allowed actually to make any sort of observation that would directly provide the answer to the question, then I will respond that I am a physicist, not a divine, and have no interest in unanswerable questions.

In fact, I am not really as stern as this. Over the years I have developed a certain faith in my theoretical moon, the one I carry around in my head; its properties are constant, reliable, and predictable, and I can use it with absolute confidence to predict at any time where a moonlike image is to be found in the sky. I trust my theoretical moon, and so does everyone else, so if anyone asks me, "Is the Moon really there when no one is looking at it?" I respond warmly, "Sure, why not, it might as well be."

But when I turn to my theoretical electron—the mental construct, possessed of certain attributes and qualities, that I also carry around in my head—things are not so amiable. It's true that if anyone asks me a question about the behavior of an actual electron in a real experiment, I can turn to my theoretical electron and use it to make predictions. But the best predictions I can come up with are merely probabilities. If someone asks, "Will the electron's spin point up or down?" I have to say there's a fifty-fifty chance of either result; and if someone else asks, "Will it point left or right?" I have to say there's a fifty-fifty chance of that too.

And if someone asks, "Does the electron's spin point in any particular direction when no one is measuring it?" I have to say, unequivocally, no.

What's the difference between the Moon and an electron? I can't be altogether sure the Moon is there if no one is looking at it, but I can be sure (because of its utility over many long years) that my theoretical moon, constant and reliable as it is, has the same properties no matter who asks me questions about it. So I can be sure that there is a theoretical moon of utter dependability and trustworthiness.

But my theoretical electron is not nearly so independent a creature. If someone asks me whether the electron's spin is pointing up or down, I should really ask my questioner, "Are you planning to measure whether it is up or down, or were you just making a casual inquiry?" It all depends. . . .

This is what makes quantum mechanics, and in particular the Copenhagen interpretation of it, so disconcerting to physicists brought up to believe in the existence of a dependable, objective reality. Always in science, we conduct experiments, obtain data, and make inferences from the data. In classical physics you can infer, for example, a lunar orbit from observations taken over a number of nights, and this inferred orbit will correctly give the position of the Moon on any other night.

But in quantum mechanics, this doesn't work. If you have an unmeasured electron, its spin is entirely uncertain; if you measure its spin using a Stern-Gerlach magnet of some particular orientation, you will obtain a specific result; you can then say with confidence that the spin-state of the electron is, for the time being, definitely in the direction it was found to be in, and you can use that datum as a baseline to predict (with the appropriate probabilities) the result of another spin measurement in a different direction. And so on. But at each measurement the spin is reset, loosely speaking, to a new value, and loses all memory of its previous value. This would be like saying that if I observed the position of the Moon on Monday, Tuesday, and Wednesday nights, then I could predict the Tuesday position from the Monday observation, and the Wednesday position from the Tuesday observation, but not the Wednesday position from the Monday observation, because the intervening Tuesday observation would have "reset" the Moon's position and made the earlier observation irrelevant.

It was an article of faith in classical physics that all observations, taken together, refer to a single, consistent reality. Quantum mechanics disallows this certainty: a series of measurements of a single object or system does not, in general, yield a set of results that can be consistently referred to a single underlying model of what is "really" going on. It seemed to Einstein, and has seemed to others over the years, that if you take away from science the idea of a unique underlying reality that all observers can agree on, then you are taking away one of the foundations of science itself. This was why Einstein worked so hard to prove that quantum mechanics was at best an incomplete theory of the world.

The fatal blow?

These problems of interpretation, and the sort of willful blindness enforced by the Copenhagen interpretation (you're just not allowed to think about what is going on beneath the surface . . .) convinced Einstein that quantum mechanics was either wrong or incomplete. He never came up with a satisfactory alternative, but in 1935, with the help of two younger colleagues, he devised a "paradox" that, to his mind, brought forcefully into daylight a profoundly unacceptable feature of the theory. The original argument proposed by Einstein, Boris Podolsky, and Nathan Rosen went something like this.

First, imagine that you can create a pair of particles, moving off in opposite directions at the same velocity. If at some later time you measure the position of one particle to some reasonable accuracy, then you know the other particle must have traveled just as far from the source, so you know its position too, to the same accuracy.

Alternatively, you could choose to measure the momentum of one of the particles, and thereby immediately know the momentum of the other.

This simple state of affairs, according to Einstein, Podolsky, and Rosen, could not possibly make sense. Such an experiment, they argued, contradicted Heisenberg's uncertainty principle.

Wait! Did we miss something? By now we are well aware that an experimenter can choose to measure either the position or the momentum of a particle to some desired accuracy, but not both. Therefore, a physicist who has set up an Einstein-Podolsky-

Rosen (EPR) experiment can, by making measurements of the first particle, deduce the momentum and the position of the second particle only within the bounds of the uncertainty principle. The uncertainty principle remains intact, both for the first particle, whose properties are measured directly, and for the second, whose properties are deduced from the measurements of the first. What's the problem?

The "paradox" arises, in the 1935 EPR paper, because the authors kicked off their discussion with a philosophical declaration. If, they said, "without in any way disturbing a system, we can predict with certainty . . . the value of a physical quantity, then there exists an element of physical reality corresponding to this physical quantity." Einstein, remember, was an out-and-out realist; he believed that measurements made by physicists refer to an objective, definite reality, independent of the physicists' acts and intentions. For example, I may not happen to know the temperature of the surface of Pluto, but I can think of sending a spacecraft there with a thermometer, to land on the surface and radio back a measurement. Less extravagantly, I could make careful infrared and radio observations of the brightness and reflectivity of Pluto, and from such data estimate what the surface temperature must be. One way or another, I can in principle find out what Pluto's surface temperature actually is, and I have no hesitation in thinking of a real quantity called "the surface temperature of Pluto." That temperature exists, and has some specific value regardless of whether I have direct knowledge of it or not. I certainly do not think of it, Copenhagen-style, as an uncertain, ill-defined quantity that acquires a definite value only at the moment I am able to measure it. The surface temperature of Pluto is, in short, what EPR choose to call "an element of physical reality." It exists; it's out there; it's a real and trustworthy quantity.

Back to our two particles, flying away from each other and now sufficiently far apart that they are quite independent. No physical influence connects the two. So, goes the reasoning of

EPR, if by performing an experiment on particle 1 I can determine the position of particle 2 as accurately as I like, without actually doing anything to particle 2 directly, then I have to conclude that "the position of particle 2" is an element of physical reality. I can find out what it is without at all disturbing the particle itself.

Equally, I could decide to measure the momentum of particle 1 as accurately as I can, which tells me the momentum of particle 2 also without doing anything to disturb it. By the same argument, I must conclude that "the momentum of particle 2" is also an element of physical reality.

I recognize in all this that when I make a measurement on particle 1, I fatally change its state, so that I cannot subsequently measure its momentum and pretend that I am measuring the same particle whose position I just determined. For particle 1, the uncertainty principle holds in the familiar manner: once I measure one quantity, I change the thing I am measuring, and cannot then measure some other property of the same pristine state.

But for particle 2, I can in principle measure either its position or its momentum with whatever accuracy I choose, without doing anything in any way to disturb its state. According to the philosophical declaration of EPR, both the position and the momentum of particle 2 must be "elements of physical reality." That is, because the properties of particle 2 can be deduced without actually making a measurement on it, and therefore without doing anything that seems to constitute a "measurement" in the quantum mechanical sense, we have to conclude that the properties of particle 2 are real, preexisting, predetermined properties. They are, therefore, classical rather than quantum properties, with a meaning and reality independent of whether or how they are measured.

Furthermore, if the position and momentum of particle 2 can be precisely found out in this measurement-independent way (independent, that is, of any direct measurement of particle 2), then the product of their uncertainty can be as small as we like, and certainly smaller than the uncertainty principle allows. And smaller, of course, than the uncertainty principle would allow if we tried to measure the position and momentum of particle 2 directly.

The point made by EPR is therefore not that the position and momentum of particle 2 could in fact be determined with accuracies that would contradict the uncertainty principle; the argument rather is that because either quantity could, in principle, be measured to arbitrary accuracy without any disturbance to the particle itself, the quantities must possess what physicists (Einstein, in particular) would describe as objective reality. And this is what goes against the reasoning of quantum mechanics.

Niels Bohr's response to this was abrupt. Agreeing, of course, that no experiment could accurately and simultaneously determine the position and momentum of particle 2, he observed that in any practical sense the uncertainty principle holds fast: it proclaims that you cannot simultaneously determine the position and momentum of a particle with unlimited precision, and this stricture is just as true for particle 2 as it is for particle 1. So, no problem. No real, empirical, or practical problem, at any rate.

One does not have to be altogether obtuse to suspect that a conceptual problem, at least, remains. Bohr's argument amounts, at bottom, to saying that you cannot legitimately combine or compare the results of two experiments that cannot both be done. In other words, it's invalid even to notionally join together, as EPR did, the results of one experiment in which you determine the position of particle 2 by measuring the position of particle 1, with the results of another experiment in which you do the same thing for the momentum of particle 2. The argument fails because these two experiments are in fact mutually exclusive. If you do one, you cannot possibly do another, and vice versa.

Thus stated, the principle seems quite reasonable. Nevertheless, it denies to physicists a freedom they have often asserted in assuming the definiteness of a physical quantity—the surface temperature of Pluto, for example—even if they did not have the actual number at hand and could not easily do an experiment to find it out. In classical physics, quantities were always assumed to be definite whether we measured them or not; in quantum physics, as we say from the beginning, that notion has to be abandoned. Bohr's response to EPR merely restates that dictum more forcefully than before.

Bohr's response, fair-minded but stern as it is, contradicts the core of Einstein's belief about the nature and meaning of "elements of physical reality." Taking Bohr seriously means that we are not permitted to believe that particle 2 has a definite momentum or position simply because we have deduced the quantity from a corresponding measurement of particle 1. More to the point, Bohr's view seems to imply, if the position and momentum of both particles are genuinely undetermined until a measurement is made, that any measurement of particle 1 that reduces its indefiniteness to a definite number (collapsing its wavefunction, if you like) also reduces the indefiniteness of particle 2, collapsing its wavefunction at the same time, even though the two particles are by now far apart. This is what happened right at the beginning of our story, when opening a box in Los Angeles to reveal a right-handed glove apparently had the instantaneous effect of turning the magical indefinite glove in Hong Kong into a left-handed one.

We seem to have a dilemma. If we take Einstein's side, we have to conclude that both particles really have, in a genuine, old-fashioned, classical kind of way, specific values of position and momentum even when those quantities have not been measured. But that seems to contradict everything we have learned so far about the nature of quantum measurement (as in a Stern-Gerlach magnet, for example) and about the uncertainty principle.

But if we take Bohr's side, and insist that all quantities are indefinite and meaningless until measured, then we seem to be saying that measurement of one member of a linked pair of particles has an instantaneous effect on its partner. This becomes problematic because in quantum mechanics, unlike classical physics, measurement is not simply the passive ascertainment of a preexisting property, but the production of a definite datum through the active involvement of measurer and thing measured. And in the EPR experiment, we seem to have a measurement that can be accomplished remotely, without any apparent connection between measurer and thing measured. This is the "spooky action-at-a-distance" to which Einstein took exception.

It seems we are forced to choose between two positions, neither of which is very appealing.

A new spin on the puzzle

David Bohm, an American theoretical physicist, began his career in quantum mechanics with doubts about the Copenhagen interpretation. To assuage those doubts, he studied the ideas of Bohr and others and wrote a highly regarded textbook on the subject, presenting the Copenhagen view as soundly as he could. Nevertheless, he was not entirely persuaded even by his own arguments, and on discussing the book in 1951 with Einstein, Bohm found his own doubts resurfacing. After all this effort to assimilate the Copenhagen philosophy, Bohm then spent much of the rest of his life trying to patch up what he and Einstein agreed were the deficiencies of standard quantum mechanics.

In the course of writing his book, Bohm came up with a version of the EPR "paradox" that more clearly brings out the puzzling circumstance at its core. For the essential pair of particles, imagine now a pair of electrons, moving in different directions, whose total spin you know to be zero. This, like

EPR's original presentation, is what physicists have taken to calling a "thought experiment"—something that could be done in principle but may in practice be dauntingly difficult. But it's possible to imagine how two such electrons may be obtained: there's a certain subatomic particle called the neutral pion, which can occasionally decay into two electrons. The pion has zero spin, and since the laws of quantum as well as classical physics dictate that spin is conserved, the two electrons that emerge must together also have zero spin.

Now, you set up detectors some distance from the source of these two electrons, but this time the detectors will not measure position or momentum. Instead, they are Stern-Gerlach magnets, which will measure the spin of the electron passing through. Here comes the "paradoxical" part: it is up to you to choose how the Stern-Gerlach magnet is oriented—vertically, horizontally, or anything in between. But choosing that direction defines the axis against which the electron spin must be referred, so that if the Stern-Gerlach orientation is vertical, then the measured electron must have its spin up or down. And if you measure the first electron to be "up," for example, you must immediately conclude that the now distant second is "down," even though you also know, from the way Stern-Gerlach devices work, that the spin of both electrons is supposed to be indefinite until a measurement is made. It seems that a spin measurement on one electron has in effect "measured" the spin of the second, remote electron, without any actual physical intervention. The second electron has been reduced from a wholly indefinite unmeasured state to a definite "down" state, even though nothing has been directly done to it. (See Figure 6.)

This is like our original scenario with the two magical gloves: you find out that one is right-handed, and deduce that the second has been forced to become left-handed.

But there's more to it than that. In the original EPR formulation, an experimenter could choose to measure either the

FIGURE 6

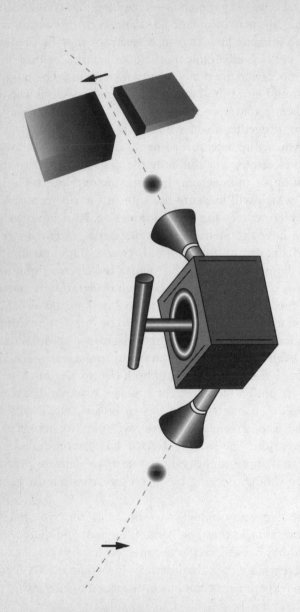

In the Einstein-Podolsky-Rosen (EPR) experiment with a pair of electrons whose spins are known to add to zero, it seems that both spins are indeterminate until a measurement is made, but that as soon as the first electron's spin is measured, the second's spin becomes definite although it has not itself been measured.

position or the momentum of the nearby particle, thus determining the corresponding property in the other. That these two different measurements can be done implies, because of EPR's philosophical predilection about what constitutes an "element of physical reality," that the position and the momentum of both particles are somehow real, not uncertain, before being measured.

In the new version, with electron spins, we can substitute for position and momentum two spin measurements that are also linked by an uncertainty principle—namely, the spin in a vertical direction (up or down) and the spin in a horizontal direction (left or right). You can measure one or the other of these quantities, but not both.

This reinforces the idea, which follows from Bohr's view of the argument, that measuring the spin of one electron instantaneously, and in some physically meaningful way, determines the spin of the other. If the electrons do not know in advance in which direction their spin is going to be measured—up or down, left or right—then it seems they must retain their full quantum indeterminacy right up to the moment of measurement, and that there must then be an instantaneous "collapse of the wavefunction" for both electrons, once the spin of either one is measured.

On the other hand, it's equally hard to know how to interpret this experiment from the point of view of Einstein and his beliefs about elements of physical reality. The fact that a spin measurement of one electron fixes, without disturbance, the spin of the other would force a follower of Einstein to conclude that electron spin must constitute precisely such an element. It must be, in other words, an intrinsic, objective property of the electron.

But this seems impossible to reconcile with the idea, supported by long experience with Stern-Gerlach magnets, that spin *must* be indeterminate until measured. How can spin, which has to be defined with respect to some axis or direction, possibly be an element of physical reality when you don't even know, at the outset of the electron's flight, in which direction the spin will eventually be measured?

Technically, it makes no real difference whether you formulate the EPR argument in terms of position and momentum or in terms of spin. But for most people, the spin formulation has a starker impact, because of the absolute nature of the uncertainty principle for spin. Position and momentum are compromised by the uncertainty principle, in that the more you know about one, the less you can know about the other; nevertheless, you can always measure both position and momentum to at least some degree of accuracy. With spins, however, the uncertainty principle is blunt: once spin is measured in one direction, its value in the perpendicular direction becomes altogether meaningless. And the dilemma posed by the EPR argument becomes inescapable.

In which Einstein is caught in a self-contradiction

However the EPR argument is presented, it was intended by its originators to show that the notion of "elements of physical reality," which Einstein thought both self-evident and central to the success of the physicists' view of the world, came into direct conflict with Bohr's view of the workings of quantum mechanics. And yet the EPR proposal does not, in the end, sit so easily with Einstein's idea of physical reality either. A slight variant of the EPR argument brings this point out with particular clarity.

As in Bohm's version, let two electrons whose spins add to zero set off in opposite directions. The first runs into a vertical Stern-Gerlach magnet, which of course measures its spin to be either up or down. The second Stern-Gerlach device, twice as far from the source of the electrons as the first, measures the spin of the second electron, but does so against a horizontal axis, yielding either a left or a right result.

Imagine that Einstein is at the first Stern-Gerlach magnet, and finds that the first electron comes out with spin up. He will conclude that the second electron must have its spin down and, told that this electron will in due course pass through a horizontal Stern-Gerlach magnet, he will predict it must come out left or right with fifty-fifty probability, since an electron that's definitely "down" is equivalently in a "half-left, half-right" state with respect to a horizontal measurement. This will be a correct prediction.

But what if Einstein is at the second, more distant detector instead? He sees the second electron come through the magnet, and he sees it come out in the beam corresponding to a left-directed spin. What will he deduce from this? If he does not yet know what the result of the measurement at the first magnet was, he can only deduce, since he has found a "left" electron, that its partner was "right," and that as it went through the first magnet it would have had a fifty-fifty chance of coming out either up or down. This too would be a correct deduction or, to use a word scientists sometimes employ, a correct "retrodiction" of what happened at the first magnet.

No matter which detector Einstein watches, he can use his observations to make a correct inference about what will happen, or did happen, at the other detector. So far, so good.

But what about the "elements of physical reality" in this case? The first Einstein will conclude that the second electron must be down, because of the up measurement at the first magnet. And because this statement is made about an electron distant from the detector, and which has not been tampered with in any way, it qualifies as an element of physical reality. By the EPR philosophy, the second electron's spin "really is" down.

The second Einstein, on the other hand, makes a deduction about the prior state of the other electron, rather than a prediction about its future. Finding the second electron's spin to be left, he would deduce that the other electron's spin must have been right, and again, because this statement is made about a distant electron which has not been tampered with (by the second physicist) in any way, it also qualifies as an element of physical reality. To the second Einstein, the first electron's spin "really was" right.

But if we put these two deductions together, we discover that the two Einsteins can, in a single experiment, deduce two elements of physical reality—the second electron is down, the first was right—which, however, are inconsistent with each other. The rules of the particular experiment require absolutely that the total spin of both electrons must be zero, so that it is impossible to have one that is "really" down and another that is "really" right. Those values do not add up to zero.

In this example, therefore, it seems that what are supposed to be incontrovertible "elements of physical reality"—meaning, in the EPR philosophy, objective and unarguably true facts—can turn out to be mutually contradictory. Which leads us to think that perhaps this idea of "elements of physical reality" is not so self-evident after all.

You might object that the first Einstein, who makes a measurement before the second electron gets to the second magnet, is in a privileged position, and is therefore entitled to some sort of precedence over the second Einstein in defining the true elements of physical reality. But this is a false distinction. As both a practical and a logical matter, deducing the past from a given set of data is really no different from predicting the future: in both cases, Einstein applies the same laws of physics and the same deductive philosophy to the same set of data, and the only difference is the entirely arbitrary one of whether you are more interested in what will happen or has happened at the other detector. An astronomer might wish to use current observations of the solar system to estimate how long it will survive; a geophysicist might use the same data to figure out how long ago it was born. Both are laudable scientific goals, and are logically indistinguishable.

In the same way, the second Einstein, even though he may know that a first measurement has already been done, is just as entitled to a view of "physical reality" as the first, and uses the same logical arguments to deduce from the available data what that physical reality must be, or must have been.

The young Einstein, who was a fanatic for paying attention to empirical facts and avoiding "obvious" but unjustified assumptions, would immediately have concluded that quantum mechanics does not permit the assumed but unproven EPR philosophy of what constitutes an element of physical reality. But the older and less flexible Einstein insisted that his notion of physical reality must be right, and that something about quantum mechanics must be wrong.

Whose reality is the real reality?

If two observers of the same experiment can hold contradictory but equally legitimate views of what is really going on, are we not led into a form of relativism, in which no point of view has absolute standing, and anyone's point of view has equal validity? Some casual philosophers and thinkers seem to believe that if quantum mechanics is supposed to underlie all the rest of physics, and that if physics above all varieties of human thought is supposed to be solidly objective, then the apparent denial in quantum mechanics of the existence of any satisfactorily definable objective reality means that no objective understanding of the world can be possible. Ultimately, so we are invited to think, quantum mechanics lends support to extreme forms of philosophical and even moral relativism: there is no absolute truth; no one person's viewpoint is finally superior to any other. And this kind of loose thinking, this abandonment of self-evident classical right and wrong, is the undoubted cause, as numerous American politicians have declared, of teenage pregnancy, inner-city decay, and the trade deficit with Japan. Is quantum mechanics to blame for the deteriorating quality of modern life?

Not so fast. Quantum mechanics indeed permits—in fact, demands—a kind of relativism, in the sense that different observers using the same logic can make different inferences about the "reality" that lies behind a set of experimental data. But this relativism is of a rigidly proscribed kind. It does not mean, in the EPR example we just looked at, that the two Einsteins can make any inference they like about the two electrons in the experiment. They both agree that there are, indeed, two electrons, they agree on where the electrons came from and where they are going, they agree on the spin measurements made on the two electrons as they pass through the separate Stern-Gerlach devices. The only point on which they are permitted honorable disagreement concerns the inferred spin state of the electrons in the interval between the two measurements.

And this disagreement, let us emphasize once again, concerns not any known or definite facts, but only an inference about an intermediate state that is never actually observed. How this calls into question the traditional physicists' view of reality is a deep and complex issue that we will have to dig into. Regardless of all that, there is nothing here about morality and politics.

In which Niels Bohr is obscure, even by his own standards

The EPR argument undoubtedly causes trouble for anyone's view of reality, Einstein's included. How are we to understand it by means of the Copenhagen philosophy? Niels Bohr thought deeply about its implications, as he did about the meaning of quantum mechanics in general, and as the father figure of the Copenhagen interpretation his writings on such matters have taken on a sort of talmudic importance, playing the role, so to speak, of úr-documents to be consulted even now in times of crisis when difficult questions are to be resolved. This is not altogether a compliment to Bohr: although one may well have the impression that, by powerful cogitation, Bohr achieved a personal understanding of the significance and correct interpretation of the EPR argument, it's something else again whether he was able to convey this understanding to others. If modern physicists still find themselves unable to explain clearly what is going on in the EPR argument, and resort instead to the delphic words of Bohr to throw at their critics, it's perhaps as much to shut the critics up by force of authority than to achieve illumination and agreement.

Be that as it may, here's what Bohr had to say, eventually, about EPR, in particular about their definition of reality as that which can be determined or deduced without disturbance to the system in question:

> There is essentially the question of an influence on the very conditions which define the possible types of predictions regarding the future behavior of the system. . . . This description may be characterized as rational utilization of all possibilities of unambiguous interpretation of measurements, compatible with the finite and uncontrollable interaction between the objects and measuring instruments in the field of quantum theory.

The first part of this quotation is often repeated, with little or no further explanation, as the definitive statement from Bohr on how the EPR "paradox" is to be escaped. The Irish physicist John Bell, who is about to enter our story in an important way, had the courage to admit in 1981 that he had "very little idea what this means."

Bohr's views of quantum mechanics are indeed remarkable in the extent to which his advice on interpretation has turned out to be both useful and justifiable, even though the technical justification may not have come for many years after the advice. Bohr evidently had a feel for the right way to proceed—by all means the mark of a great scientist. Even so, it is a curious historical episode in science that so much effort has been devoted to teasing scraps of meaning from Bohr's ponderous and sometimes incomprehensible remarks.

For many years, physicists and philosophers talked about the EPR "paradox" (never quite deciding if there actually was a paradox, and if so, what precisely could be said to be paradoxical), not to mention Schrödinger's cat, Wigner's friend, and the old reliable two-slit interference experiment. This was almost entirely an abstract discussion, running from debate over the nature of reality, the meaning of "measurement," the demarcation of the boundary between quantum systems and the apparently classical devices that must be used to generate definite results from indeterminate states, the possible connection between physics and consciousness, the existence or not of influences (directly detectable or not) that seemed to instantaneously affect distant particles, and so on. Nothing much was resolved for a long time, and many physicists stood deliberately by, perhaps intrigued by the "meaning" of quantum mechanics but at the same time unclear what exactly was being debated and happy in any case to know that quantum mechanics seemed to work unambiguously, whether anyone really knew how or not.

Scientific progress, by long-standing convention, is supposed to come about through experiments that can test competing hypotheses by producing a decisive result in favor of one theory, and in contradiction to the rest. In quantum mechanics, what we seem to have is a single theory that generates undisputed predictions, along with a nebulous and free-floating debate over what those predictions "mean." For many years, there was indeed nothing that could be put to the experimental test, but through the debate over meaning a better understanding of the fundamental issues slowly dawned, and a couple of revisionist views of quantum mechanics attracted small bands of enthusiasts.

And how many universes did you say you'd be needing?

The Copenhagen interpretation of quantum mechanics boils down to these two ingredients:

First, a quantum system that hasn't yet been measured exists in a state of genuine indeterminacy. It makes no sense (and may actually lead to contradiction) to say that it is in a specific but unknown state. Second, the act of measurement forces the system to adopt one of the possible states allowed to it, with a probability that can be calculated from the appropriate wavefunction for that system and that measurement upon it.

The first part of this interpretation denies the existence of an independent objective reality of the sort that physicists had come to know and love. The second relies on a seeming act of magic, since neither Bohr nor anyone else was ever able to explain what constituted an "act of measurement" or how such an act made the indeterminacy go away.

Many physicists therefore found the Copenhagen interpretation unsatisfactory, despite its unfailing practical utility. But what alternatives could there be? What views of physical reality might be in competition here?

If the act of measurement is, in Bohr's interpretation, a piece of inexplicable magic, one way to avoid it is to argue that measurements never actually happen. In 1957, Hugh Everett, then at Princeton University, made a novel suggestion. When a measurement is made, it seems that only one of the possible outcomes of that measurement takes on a genuine reality, and that the others, the outcomes that might have occurred but don't, simply vanish. Everett proposed that the nonoccurring outcomes are not lost at all, but carry on a parallel existence in other worlds.

Loosely speaking, we are supposed to think that, before measurement, the possible outcomes coexist, but that the effect of measurement is to force each possible outcome to carry on a separate, noncommunicating existence in its own universe. Each separate universe must contain the measuring apparatus, of course, and if that apparatus includes you or me, operating the controls and waiting for the result, then we have to conclude that "copies" of ourselves also propagate in their separate ways, in these separate universes.

The main point of the Everett interpretation is that if we allow ourselves a lordly view in which we imagine all these distinct universes going their separate ways, then we do not have to worry about the disappearance of any of the possible outcomes of the measurement. In each universe, a measuring device, and perhaps a copy of you or me to watch over it, continues along, one for each possible outcome. Every possible result of the measurement occurs in one of these universes, and nothing is lost.

This sounds somewhat metaphorical. It is most definitely an interpretation of quantum mechanics, not a distinct version of it. Although the idea of separate universes, each one carrying a possible version of what might have happened, appeals to some people, Everett's suggestion has never found great support among physicists. One of the key points of standard quantum mechanics is that although possible outcomes can interact with each other before a measurement (as in, for example, the interference between two different photon paths in a two-slit experiment), they must be scrupulously independent after measurement (so that if you measure the photon before it has a chance to get to the screen where the interference pattern is revealed, the two possibilities become physically distinct, and no interference occurs). Everett's many universes must be rigorously and absolutely separate; any later interaction or interference between universes containing different outcomes of the same measurement would constitute a violation of quantum mechanics. It would be like finding a way to see the interference pattern in a two-slit experiment even after you have made a "which way" measurement that forces a distinction between the two possible photon paths corresponding to the two slits.

But if, by definition, the separate universes are strictly noninteracting, there's no possibility of doing an experiment in one universe that would reveal the existence of another—which seems to render the fundamental point of Everett's proposal neatly immune to any possible test. And when you think about how many of these parallel universes you have to provide, the whole idea begins to seem cumbersome, to say the least.

If you are thinking only of measuring the spin of a single electron or the polarization of a single photon, you need only two universes, one for each of the two possible outcomes. Similarly, in the simplest kind of EPR experiment, in which the spin of a pair of electrons is measured by two vertical Stern-Gerlach magnets, we simply imagine that when one spin is measured, a split into two universes occurs such that, in one universe, the spin we measure is up and the unmeasured one is down, and in the other universe, the opposite holds. Both outcomes occur, each in its own universe.

In a more elaborate EPR experiment, where the spin of the first electron was measured by a vertical magnet, and the spin of the second by a horizontal magnet, at some later time, the splitting into universes gets a little more complicated. The first spin can be up or down, and for either of these two possibilities, the second spin can be left or right. So now you need four universes. Still, this might seem not unmanageable: in any series of experiments, universes keep splitting off, but at each turn only a finite degree of multiplication occurs. It may get hard to keep track of all these proliferating universes, but nothing conceptually new occurs when we perform increasingly complicated experiments, or just more of them.

But there are other kinds of measurement that are not so tidily defined, and which cause some deeper problems for the many universes interpretation. How are we to think, for example, of Compton scattering, in which an electron and a photon collide? In that case, the number of possible outcomes seems to be infinite for this single event alone: if the collision is characterized by the angle through which the photon's initial path is diverted, then 45 degrees is a possibility, but so is 45.1, and 45.11, and 45.111 . . . ad infinitum. Do we really have to imagine that one parallel world is born for every decimal place in the angle?

Even worse, think of what happens when a photon strikes your eye, letting you know that the sky is blue. That photon has battled its way from the center of the Sun to your retina, colliding along the way with innumerable (well, OK, not strictly innumerable, but very many) atoms and electrons: in the Sun, between the Sun and the Earth, in the Earth's atmosphere. Each one of these trillions of events is Compton scattering or something closely related, yet the only thing you actually register is the photon finally initiating a biochemical reaction in your eyeball, which sends an electrical signal to your brain signifying that light has been seen by you.

The number of possible paths that photon could have taken in getting from the center of the Sun to you is literally infinite. Any number of possible collisions and trajectories could have delivered the photon from the center of the Sun to your eye, and along every such path, each individual scattering event has to be accounted for with the appropriate splitting of universes. It then appears that Everett's proposal requires the generation of an infinite number of possible universes for a whole string of more or less inconsequential but nonetheless quantum mechanical events.

If you're an enthusiast for the many worlds idea, this kind of criticism is empty: once you've taken to heart the idea of lots of parallel but unseen and unseeable universes, quibbling whether you need a lot or an infinite number is perhaps beside the point. Because the many worlds interpretation is standard quantum mechanics with an unusual metaphysical gloss, it can never be proved or disproved, and it is guaranteed to account for anything within the compass of quantum mechanics. But, as has been wisely said, a theory that explains everything explains nothing.

And what, in any case, has been achieved? The problem with Copenhagen is that it leaves measurement unexplained; how does a measurement select one outcome from many? Everett's proposal keeps all outcomes alive, but this simply substitutes one problem for another: how does a measurement split apart parallel outcomes that were previously in intimate contact? In neither case is the physical mechanism of measurement accounted for; both employ sleight of hand at the crucial moment.

For these reasons Everett's "many universes" proposal has never really got much beyond the mere statement of an idea. And in favor of Copenhagen, there's something to be said for not creating more universes than one really plans to live in.

Indeterminacy as illusion

If you want to believe that we live in only one universe after all, the way to avoid the unsettling implications of the Copenhagen interpretation is to deny that the indeterminacy of quantum mechanics is real. This was Einstein's preference. In the classical world, indeterminacy always means ignorance: if you knew precisely the trajectory of a spinning coin, you could say whether it was going to land heads or tails. Likewise, Einstein believed, if you knew more about quantum systems than Bohr said you could know, you would be able to predict which particular outcome of any measurement was actually going to emerge. If some deeper theory were eventually to "explain" quantum mechanics in terms of still more fundamental precepts, we would see the indeterminacy of quantum mechanics simply as the consequence of our ignorance of the true nature of the world.

Restoring order in this fashion may seem like a noble aspiration, but what we know of quantum mechanics so far should tell us that constructing such a theory is not going to be easy. The two-slit experiment, in particular, seems to require a genuine indeterminacy as to the slit that a photon goes through, if an interference pattern is to be created. But the desire for determinism is strong, and physicists are ingenious. A particularly ingenious quantum physicist was David Bohm, already notable for his more comprehensible rendition of the EPR proposal. Bohm began as a traditionalist, hewing more or less to Bohr's ideas on what was to be made of the predictions of quantum mechanics, but he came to find Bohr's way of thinking strange and restrictive (as indeed it is). After many years he invented a workable version of what has become generally known as a "hidden variables" kind of quantum mechanics, in which measurements seem to follow probabilistic laws only because we are ignorant of certain secret or "hidden" properties of the things we are measuring. If we could know the values of the hidden variables, we could say precisely what outcome a measurement would produce.

In essence what Bohm did was to give specific and almost classical meanings to the "wave" and "particle" halves of quantum mechanics. Rather than saying, as Bohr tried to, that the wavefunction was what one used to calculate the internal probabilities inherent in some experimental setup, and that the particle only became apparent when a measurement was made and some definite outcome obtained, Bohm asserted that the wave and the particle were both always present.

In a two-slit experiment, for example, Bohm says that a photon does indeed exist, and must, because it is a particle, go through one slit or the other. But he adds that a wave also exists, spreading throughout space and negotiating, as only a wave can, both slits at once. The new ingredient is that the wave influences—in fact, determines—the photon's path; it is called for this reason a guide wave or a pilot wave.

In other words, the guide wave propagates throughout a double-slit apparatus, spreading through both slits, and creating, on the far side of the slits, a classical interference pattern. Individual photons may then go through one slit or the other; it doesn't really matter, because the guide wave, as its name implies, guides the photons' paths, leading them to strike the screen so as to build up the appropriate striped pattern of light and dark.

The theory is deterministic, in the sense that the guide wave lays down a fixed pattern in space, and any individual photon's motion is controlled without variation or randomness by this spatial pattern.

Now here's a problem: if this is really what's going on, then a stream of identical photons going towards the two-slit apparatus would all end up in the same place on the screen. It would be like a golf ball rolling across an undulating green: the track that the golf ball would follow might be complicated, but it would be predictable, and if you could hit the golf ball every time with precisely the same speed and direction, it would always follow the same track and come to rest in the same place. If this is what happened in a two-slit experiment, in Bohm's interpretation, we would see a bright spot at some place on the screen, not a general pattern of light and dark stripes.

So a second ingredient is needed to make this theory work. In standard quantum theory, we regard a stream of photons as having nominally identical momentum and direction, but we say that the uncertainty principle makes it impossible to define with absolute precision the exact momentum and direction of any single photon; all the photons, you might say, are as identical as we can physically make them. In Bohm's version, on the other hand, each photon has, with full exactitude, a certain momentum and direction, but a stream of photons that we can manufacture always has some residual variation within it, so that no two photons are precisely identical. In this way we are supposed to think of the photons as we might think of the atoms in a volume of gas; the average speed of each atom is the same, but at any one time, some are moving a little faster and some a little slower.

So it must be with the photons heading towards the two slits. Because they are not moving all exactly in identical fashion, some go through one slit, some go through the other, and once past the slits the individual variation is enough to send them all on slightly different paths through the guide wave field, arriving in different places at the screen and thus creating the requisite interference pattern. It is as if, as would in practice happen on the putting green, you hit a series of golf balls with more or less the same speed and direction, but with some unavoidable variability. Each ball would then follow a slightly different track, some going a little farther, some not so far, some to the left, some to the right, so that if you hit a hundred such golf balls they would come to rest in a pattern that reflected the undulations of the grass they had just traveled across. This is how Bohm's theory explains interference in the two-slit experiment: the guide wave pattern acts as the surface of the putting green, sending any photon that travels across it in a predictable path; but the photons all have slightly different initial motion, so they end up in different places, to create a pattern rather than a single bright spot.

The theory, in effect, builds quantum mechanics onto a classical foundation. Everything is deterministic. Every photon follows a predictable path, and it is only because you do not have a precise knowledge of the motion of each photon that you have to resort to probability to describe where they will end up.

The precise but unknown momentum and direction of each photon constitute the "hidden variables" of the theory. If you knew what they were, you could predict each photon's path with absolute accuracy. So, as in the classical theory of atoms in gases, probability is simply a result of ignorance, and is not a fundamental fact of nature. This is the kind of thing Einstein had always hoped for. Sounds appealing, does it not?

So why have physicists in general not embraced the Bohm theory?

In which seeming virtues are displayed as faults

For one thing, the way that the "guide wave" tells the photons where to go is quite mysterious. Although it is adapted from the conventional wavefunction, it behaves in an entirely new way. What Bohm came up with by playing around with the mathematical form of the wavefunction was a classical-looking equation for the motion of a classical-looking particle, with one addition: determining the particle's motion, in addition to the classical-looking forces, was something Bohm called the "quantum potential." It's this quantum potential that gives rise to the guide wave in Bohm's theory, and a very nonclassical thing it is.

As indeed it has to be. Quantum mechanics works in distinctly novel ways, and it's not possible that merely by playing around with the mathematical form of the equations, those novelties can be made to go away. What Bohm achieved was a new formulation of quantum mechanics that is mathematically the same as the standard theory, but which is rearranged so that everything looks classical except for this one strange "quantum potential"—and that's where all the nonclassical aspects of quantum theory end up.

How strange is the quantum potential and the guide wave it

generates? For one thing, it apparently moves a photon (or any other particle) around without exerting anything we would recognize as a force. In the two-slit experiment, the guide wave is supposed to guide the photon's path so that it ends up in the right place; you can think of it as an undulating surface which has the power to move the photon this way or that to the appropriate destination. But in classical physics, any time you move a particle, you have to exert a force on it. As Newton realized three centuries ago, forces are reciprocal; there is the famous "equal and opposite reaction" of his third law. If something exerts a force on something else, the necessary application of energy entails a reciprocal change of energy in the thing being forced. When you push a car, you exert a force, you expend energy, and some of that energy is transferred to the car.

In short, if the guide wave of Bohm's theory were a classical wave exerting some sort of classical force, it would alter the energy of the photon. This simply doesn't happen. Photons arrive at the screen of a two-slit experiment, having traveled through empty space, with as much energy as they had in the first place.

But if the guide wave cannot exert a force, how do photons respond to it? Hard to say. Most recently, Bohm has suggested that in some way the guide wave carries information about where a photon should go, and that the photon is not some dumb, inert particle but can receive and "interpret" this guiding information and move accordingly, somewhat in the way that a ship's captain can receive radio directions and steer his vessel accordingly. This is a large freight of interpretation to place on a simple mathematical function.

The other distinctly nonclassical aspect of the quantum potential and the guide wave is that they have to be "nonlo-cal"—that is, the information that they mysteriously carry and transmit to the photon must be gathered instantaneously from all parts of the experimental setup. Something of this sort is unavoidable. The mere fact that an interference pattern has been created indicates necessarily that influences from different places have met and interacted. So if you want to insist that the photon behaves like a simple classical particle, responding only to the quantum potential and guide wave at its specific location, then you have to conclude that the guide wave itself carries information from every part of the apparatus. The guide wave must in effect explore all parts of an apparatus at once, so as to be able to relay the necessary information to the particle.

The nonlocality of the quantum potential shows up even more clearly when Bohm's theory is applied to the EPR experiment. Before we can understand that, however, we have to understand how Bohm explains spin, and that turns out to be a big problem in itself.

We were careful, some time ago, to disavow any too precise a connection between quantum mechanical "spin"—the thing that comes out up or down in a Stern-Gerlach measurement—and the old-fashioned classical idea of particles as little spinning tops. But the whole idea behind Bohm's philosophy is to restore to particles fixed and tangible properties. If you want to think of an electron as having, in a literal way, spin, so that it travels through a Stern-Gerlach magnet like a little spinning top, then you have to conclude that the interaction between spin and magnetic field is such that electrons coming into a Stern-Gerlach magnet with random orientations have their spins twisted around, so that half the initial alignments come out up and the other half come out down.

And the thing that causes this twisting around has to be, of course, the quantum potential, exerting its mysterious influence on the axis of the spinning top. (The shape of the quantum potential, in this case, includes an influence from the magnetic field of the Stern-Gerlach device, so that in effect the quantum potential tells the electron how to move in response to the field.)

The "hidden variables" aspect of this would be that if you knew precisely the initial alignment of any one of the electrons, you could predict whether it would come out up or down: the quantum potential acts on the spin in a wholly predictable way. But (and it has to be this way since Bohm's theory is in the end equivalent to the standard interpretation) you can never find out for sure what that initial alignment is, because to do so you would have to perform a prior Stern-Gerlach measurement, and that would twist the electron's spin around to produce the appropriate up or down answer. The electron spin has, in this view, some specific prior value, but you can't ever get at it. That's why it's called a hidden variable.

Now when you think of this sort of spin measurement on two separate electrons in an EPR experiment, you can see that when, in one magnet, the quantum potential twists the spin around so that the electron is made to be "up," then it must at the same time twist the other electron's spin around so that it becomes literally "down." The quantum potential must instantaneously transform the spin measurement on one electron into a physical effect on the spin of the other.

And this instantaneity raises yet another concern. Although it is guaranteed to reproduce precisely the predictions of conventional quantum mechanics, Bohm's theory has a flaw: it's "nonrelativistic"—meaning that it embodies a notion of time and space derived from Newtonian physics rather than from Einstein's relativity.

The problem is that although Bohm must accept the existence of an instantaneous physical effect transmitted via the quantum potential from one electron to the other, "instantaneity" in relativity is not a well-defined concept. According to Newton, time is universal and immutable, and two simultaneous events are by definition simultaneous to everyone.

But according to Einstein, events happening in different places may be perceived as instantaneous by one observer, but to someone else passing by the scene at some speed, the events may happen at perceptibly different times, and even in different orders. This happens because the measurement of time as it applies to distant events is inextricably tied up with the transmission of signals to and from that event, and whereas in Newton's world light signals can travel instantly from one place to another, in Einstein's universe, nothing can go faster than light, so that any conceivable signal must take time to get from one place to another.

If we imagine an EPR experiment being performed on the grand scale somewhere in space, with observers cruising around in rocket ships at various speeds and watching the proceedings, we realize that some observers may see a spin measurement being made on electron A before a measurement is made on electron B, while others will see the sequence of measurements reversed.

But then if there's a direct physical influence telling the second electron what to do when the first has been measured, we can't say for sure which way round the influence must travel. This is no problem for the Copenhagen interpretation, because the only thing that matters is that the results of the measurements are correlated in a certain way, and we sternly resist the temptation to assume that some influence passes from one electron to the other. As long as we forcibly put aside any attempt to assign a tangible physical meaning to the wavefunction or to the "collapse" of the wavefunction that happens when a measurement occurs, it doesn't matter that one measurement precedes or follows the other.

But in Bohm's theory the instantaneous transmission of influence from one electron to the other is supposed to be a real physical phenomenon, with real physical consequences: when one electron is measured, it literally creates an interaction with the quantum potential that affects the other electron. But how then are we to reconcile this idea with the fact that some observers will decide the influence is passing from electron A to electron B, while others will decide it is going the other way? Bohm's theory insists on an objectively real description of events that's the same for everyone, but relativity doesn't allow such a definition of instantaneity.

Whether the theory can be saved from this dilemma is unresolved. No one has come up with a version of the theory that can handle Einsteinian relative time, but on the other hand no one has proved (despite suspicions that the dilemma may be fundamental) that such a theory cannot be constructed.

What does determinism mean anyway?

Still, you might think that the inelegances and contrivances of Bohm's theory are a price worth paying for the restoration of determinism. Or you might think that Bohm's work merely demonstrates that a hidden variables theory is possible, and that someone else will invent a more appealing version in due course.

On the other hand, you might be wondering about the value of a supposedly deterministic theory if in practice we are still dealing with experiments whose outcomes have to be described by probabilities, because the determinism is wrapped up in the hidden variables, which remain just that—hidden.

That's a pertinent worry. According to Bohm, every particle has definite properties, but you don't quite know what those properties are exactly, and that's why you have to resort to probability to describe the results of measurements. It's also the case that Bohm's theory is a revamping of quantum mechanics that's mathematically identical to the original, so that you can never obtain from it a result that beats the predictions of the Copenhagen interpretation. More precisely, if quantum mechanics in its usual guise can never yield a measurement that gets around the uncertainty principle, then Bohm's hidden variables theory can't do the trick either. The hidden variables stay hidden.

What this seems to amount to is an argument that although Bohm's theory is fundamentally deterministic, its predictions always carry an air of probability because you can never actually discern the underlying hidden variables. Whether this is a more or less preferable way of looking at things must be left for the beholder to decide. But it does seem a little disappointing that Bohm's theory should be advertised as deterministic, and you find out only on reading the small print that it's never actually deterministic in any practical sense.

And of course, those steeped in the tradition of Bohr will say that this is all a meaningless debate; the only thing that matters is what you can measure, and if you are forbidden, in practice, from ever seeing the hidden variables and the supposedly deterministic information they carry, it becomes a matter of metaphysical taste whether you think they're there or not. And this follower of Bohr will say that the novelties of physics required to maintain a belief in these hidden variables—in particular, the existence of this strange, not-really-physical quantum potential—are too great a price to pay for the modest metaphysical comfort that Bohm arguably affords.

You can push it around, but you can't get rid of it

By now you are perhaps beginning to understand why the great silent majority of working physicists has stayed out of the debate of what quantum mechanics means and how it is to be understood. The Copenhagen interpretation is mystifying and fails to explain how measurements ever get made. The many universes idea is willfully extravagant, and still fails to explain how measurements get made. And the hidden variables version, at least in Bohm's formulation, relies on a bizarre quantum potential and in the end restores determinism in at most a metaphysical, inaccessible kind of way.

What's the point of all this? Still, we have no distinct theories that can give predictions any different from the standard theory, so still we have no way to attack any of these issues by means of experiment. Both the many worlds idea and Bohm's version of the theory are interpretations of the same underlying theory, not different theories.

What does it mean that there are "interpretations" of quantum theory? Orchestra conductors offer interpretations of symphonies; readers and critics offer interpretations of poems. In either case the raw material—the sequence of notes or words, preserved in black and white exactly as they left the composer's hand—is unambiguous and unarguable. Nevertheless, no two performances of a piece of music or readings of a poem are identical, and there is no such thing as a definitive interpretation; there's always room for a performer's individual genius to enliven the most jaded piece of art.

Similarly, the raw material of quantum mechanics—the formulas and equations, devised through the collective efforts of many physicists and preserved within the pages of numerous textbooks—is not the topic of disagreement. The theory is rigorous and exact; physicists know how to use it, and don't argue about the predictions it makes. (There have been and will still be complicated cases where it takes some time to sort out the correct application of the formulas and equations, but even then there is, everyone would agree, a correct answer, which will duly be found.)

The trouble with quantum mechanics, and the reason why there are "interpretations," is that despite decades of practical success with the theory, physicists still cannot honestly say that they know what the theory means; they can't see inside it, so to speak. In the old days of classical physics, the formulas and equations clearly referred to an independent, objective world. Particles existed, and had positions and speeds and energies that were all well defined and unambiguous. You might not know what they were, but you could be confident they were there. Quantum mechanics, on the other hand, doesn't allow this. Nothing is anything until you measure it; only then does it become a reliable, dependable something. And worse, if you try to infer, from a given set of experimental data, what is going on beneath the surface—what is "really" there, underlying the measurements—you run the risk of finding yourself contradicted by someone else's view of what is really going on, someone who has used the same sort of data and the same kind of logic to infer, quite legitimately, a different "reality." And so you conclude that reality depends on who is looking at it, or that there is no true reality at all. Either conclusion seems to upset all the traditional notions of what scientists are looking for. In short, physicists can use quantum mechanics with ease and assurance and yet not feel that they understand what is going on.

What most people, scientists included, mean by an understanding of an idea is they have an intuitive grasp of the basic principles. We do not have to be able to calculate tiny details of the behavior of atoms in a volume of hot gas to feel that we nevertheless understand how the vigorous motion of atoms manifests itself as pressure on the walls of a container. Whether we think of atoms as dried peas or slightly squishy tennis balls or even as miniature solar systems, it makes sense to imagine pressure as the result of countless tiny impacts, individually intangible but adding up to a palpable force. And the reason it makes sense is principally that we can imagine the whole process on a personal scale—we might think, for example, of hiring dozens of small children to fling tennis balls at a wooden fence, to see if they can generate enough force to knock it over—and having imagined such an easily apprehended scene we can, in our minds, shrink the process down to an invisible scale, and think of molecules of air bashing relentlessly at the skin of a balloon, providing enough collective force to keep it plump.

This sort of visualization works for most of physics. The principles we commonly use, consciously or unconsciously, to understand cause and effect in the world around us can be translated more or less directly into principles that illuminate the microscopic world. In many contexts it is appropriate and helpful to think of atoms and other particles as little objects following the same laws of mechanics that hold for tennis balls and planets. But when we contemplate these same particles as individual quantum objects, as in the two-slit experiment, we emphatically cannot hold on to the same comfortable images. We can say to ourselves that photons behave neither like waves nor like particles, but like some intimate mixture of the two, and we can caution ourselves not to take either wave or particle imagery with too much exactitude, but we are left unable to imagine just what it is that a photon is like—for the simple reason that there is no object in the familiar world around us that behaves like a quantum object. There is no analogy to anything we can see or grasp.

And the fact that we cannot, in the end, translate the mathematically unambiguous explanations provided by quantum theory into familiar, recognizable pictures is what we mean when we say we do not, cannot, "understand" the theory.

And the resort to "interpretations" amounts to an effort to get around this difficulty, the idea being that a good interpretation of the theory would be one that made us think "ah, *now* I get it."

The difficulty, of course, is that an interpretation that makes me think I've got it may do nothing for you, and vice versa. Weird things happen in the quantum world, the two-slit experiment or the EPR kind of measurement being prime examples, and it's simply not possible to make the weirdness go away by shuffling around with the terms of the theory and calling your effort an "interpretation." The virtue of the Copenhagen view is that it takes the weirdness at face value, and declares that any effort to interpret what is "really" going on is futile. But for scientists long accustomed to making intellectual progress by digging behind mere appearances, that can be a hard philosophy to accept.

But do the many worlds or the Bohm interpretations help us "understand" any better? For some people, perhaps they do— but do not imagine that either version is any less weird than the original: one puts the weirdness in parallel unseen universes, the other in a quantum potential that moves and acts in mysterious ways. As Richard Feynman put it, "We cannot explain the mystery in the sense of 'explaining' how it works. We will just *tell* you how it works. In telling you how it works we will have told you about the basic peculiarities of quantum mechanics."

Or, to put it another way, we might imagine quantum mechanics speaking to us in the words that Samuel Johnson addressed to an obstinate critic of his views on some matter: "Sir, I have found you an explanation, but I am not obliged to find you an understanding." Although quantum mechanics provides explanations of the results of experiments, those explanations tend not, in our minds, to add up to an understanding. But why should they? It's the job of science to provide theories and models that give us an accurate picture of the way the world works, but we are not free also to demand that these theories should conform to our prior expectations of the way we would like the world to work, or think it ought to work. If science sometimes provides explanations without giving us what we would regard as an understanding, the deficiency belongs to us, not to science.

If all interpretations of quantum mechanics yield the same predictions and only try to "explain" in different ways how those predictions arise, then they are balm for the spirit, not grist for the mills of experimental physics. One might as well be democratic about it and let everyone choose, without rancor or snobbery, whichever interpretation they prefer.

So what shall we do now?

ACT II

Putting Reality to the Test

If one person made it respectable once more for physicists to think about and study the meaning of quantum mechanics, it would be John Bell. Born in Belfast, he spent most of his working life at the CERN, the European Center for Particle Physics, just outside Geneva. Whereas Niels Bohr enveloped the mysteries of quantum mechanics in a web of words as enigmatic and ambiguous as the subject he sought to explain, smothering his readers in a blanket of reassurance that warded off misunderstanding by the indirect expedient of creating a beguiling darkness that set in precisely at the point where one should want more light, and leading more than a few students of physics to conclude—not without, one may suppose, a dim sense of betrayal, for how could a loyal apprentice admit the master's thick blandishments to conceal not great wisdom but the implicit acknowledgment of failure?—that his pronouncements had begun to resemble what Thomas Hardy said was the prose of Henry James, that is, a ponderously warm manner of saying nothing in infinite sentences, Bell pursued clarity above all.

The hallmark of Bell's work was his success in locating precisely the point at which classical views of reality ran into trouble with quantum mechanics, and in devising a means by which the two viewpoints could be empirically compared. Bell's sympathies seem often to have lain with Bohm, and the effort to establish a hidden variables version of quantum mechanics. His greatest achievement, though, was to demonstrate incontrovertibly the price that must be paid to make any such theory work.

A new angle on EPR

If yesterday's weather forecast said there would be an 80 percent chance of rain this afternoon, but this afternoon has arrived and it's sunny and dry, was the forecast wrong? Not exactly: there was a 20 percent chance it wouldn't rain, so if indeed it doesn't rain we may be a little put out—we could have gone on that picnic after all—but we can hardly say the forecast was wrong. Misleading, perhaps.

And what if the television forecast had said there would be an 80 percent chance of rain, but the radio forecast had put the chance of rain at only 70 percent? Does that mean, if it actually doesn't rain, that the radio forecast was a bit more correct than the television forecast, even though both put the odds in favor of rain?

To assess the merits of a particular weather forecaster, what you would have to do is keep a scorecard: if there's a prediction of rain with 80 percent probability on twenty separate occasions, and if in fact it rains on only twelve of those twenty days—60 percent—you would have some reason to think that the weather forecaster was systematically overestimating the chances of rain, and you might reasonably conclude that the theoretical machinery that comes up with rain predictions was somehow flawed.

In one important way, testing quantum mechanics is potentially easier than testing weather forecasts, because in fundamental physics it's possible to repeat exactly the same experiment over and over. This gives a more controlled means of testing predictions that are couched in terms of probabilities; in weather forecasting, conditions vary from day to day, so an 80 percent prediction of rain on one occasion may come about for quite different reasons than an 80 percent prediction on another occasion. Even, therefore, if you find that the weather forecast is systematically wrong, it may be next to impossible to sort out, from all the different variables and unknowns that go into the predictions, what precisely is running awry. In quantum mechanics, on the other hand, you can set up identical conditions repeatedly, and so hope to control and test specific assumptions or ingredients of the theory.

In principle, anyway. Finding a way to formulate assumptions about the nature of underlying reality, and to devise experimental means under which different assumptions could be put to the test—meaning, in short, that they would lead to different probabilistic predictions over the course of a repeated series of experiments—took physicists a long time. It was only in 1964 that John Bell was able to see a way through the thicket of assumptions and interpretations and expose some of them in an experimentally testable manner. And even then it was another twenty years before the technical expertise to do the experiments became feasible.

After Einstein, Podolsky, and Rosen had formulated their "paradox," and especially after David Bohm had recast the puzzle in terms of spin measurements on a pair of electrons, physicists tended to focus their attention on two particularly simple cases.

In the archetypal example, you measure spin for both electrons in the same direction. In that case, as soon as you find one electron to be "up," for example, you know that a similar measurement on the other would reveal it to be "down." Or if you measure one to be "left," the other must be "right."

Such results lend an air of spurious definiteness to the proceedings. Because one spin measurement completely and unambiguously determines the other, it is easy to imagine, if you are so inclined, that in fact both measurements were somehow pre-ordained from the outset. The "hidden variables" idea seems ideally suited to explain such examples, because the definiteness of the results seems to suggest that the outcomes of the measurements might somehow have been implicit in the electrons' hidden properties from the start.

Alternatively, you might set the Stern-Gerlach magnets at right angles. In this case, the outcomes of the two measurements are independent. If you measure the first electron to be up, then you know the second must be down. But if you measure the second electron with a horizontal Stern-Gerlach magnet, that definite down state translates into an indeterminate "half-left, half-right" state, so that the second spin measurement has an equal chance of coming out either way—just as it would for an isolated electron that you knew nothing about. This version of an Einstein-Podolsky-Rosen (EPR) experiment doesn't seem to take you into interesting territory. It's just another example of quantum uncertainty: measure one thing, and you have complete ignorance of another.

But, John Bell imagined, what if you put your two Stern-Gerlach magnets at some intermediate angle, so that they were neither exactly aligned nor at right angles? Let's say the first electron goes through a vertical magnet, and comes out up, so that the second must be in a down state. What happens now if this down electron passes through a Stern-Gerlach magnet set at forty-five degrees from the vertical? There can be only two possible outcomes: the electron must come out in one of the two directions defined by the magnetic field, which we can call northeast and southwest. But the probabilities of these two outcomes are not equal.

That isn't very surprising. An electron in a definite down state will come out, with 100 percent probability, through the down channel of a vertical Stern-Gerlach magnet, and with fifty-fifty probability in either the left or the right channel of a horizontal magnet. For a magnet set at some intermediate angle the probabilities of the two outcomes are, naturally enough, intermediate between these two extremes. To calculate the exact answer you need to know a little more about quantum mechanics than we will go into here, but in essence it's just a matter of trigonometry. In fact, a down electron going through a magnet set on a northeast-southwest angle has about a 15 percent chance of coming out northeast and correspondingly an 85 percent chance of coming out southwest.

Bell's insight was to recognize that this is a potentially telling intermediate case. The measurement of an up state for the first electron does not tell you with certainty what the outcome of a northeast-southwest measurement on the second electron will be, but neither does it leave you with a purely random, fifty-fifty result. What we have is a measurement on the second electron that is probabilistic (since both outcomes are possible) but that is also influenced by the measurement of the first electron (since the probabilities of those two outcomes are not equal).

Still, it takes a little more work to see how this idea of misaligned magnets in an EPR experiment can be used to investigate the inner workings of quantum mechanics. Bell's ingenuity led him to a mathematical theorem which connects the outcomes of these sorts of measurements with assumptions about the "elements of physical reality," to use Einstein's phrase, and to a deeper appreciation of what information, if any, is exchanged between the two electrons in the experiment. His theorem is profound but also remarkably simple: in essence, it is a piece of high-school algebra.

Now, if the invocation of high-school algebra is enough to alarm you that something mystifying and esoteric is coming your way in the next section, let me rephrase it: Bell's theorem can be understood as long as you have the ability to add and multiply numbers. It's well worth the modest effort it takes to understand this result. How often do you get the chance to paddle in the waters on which the great physicists of this century have sailed?

Fun with algebra

To remind you of the fun that can be had from algebra, here's a little trick:

Think of a number;
multiply it by three;
add four, then multiply the whole thing by two;
add ten and divide the result by six;
now take away the number you first thought of.
What do you end up with? Why, three!

The series of operations has been ingeniously arranged so that everything cancels out to leave you with three at the end no matter what number you plugged in at the beginning. Amazing!

Back to quantum mechanics. Imagine a somewhat more complicated EPR-type experiment than we have been dealing with so far. As usual, a pair of electrons whose spins add up to zero is created by some source, and sent off in separate directions. But in each electron's path we now install two Stern-Gerlach magnets, set at different angles, and we have a switching device that can send each electron, at random, to either magnet.

In other words, the spin of each of the two electrons will be measured by one of two magnets in each path, and all four magnets are set at different angles.

To record the results, we label the four magnets A, B, C, and D. Magnets A and B are for the first electron, C and D for the second. Each magnet, when a spin measurement is made, sends an electron along one beam or the other: we label the possibilities "+1" or "−1." (If we were using a horizontal magnet, for example, we could call +1 "left" and −1 "right"). If it happens that the first electron is measured by magnet A, and comes out in the beam designated −1, and the second electron goes through magnet C, and comes out in the beam designated +1, then we would record in our laboratory notebook that this particular experiment yielded the result A = −1, C = +1. All possible results come down to some such pair of numbers.

To sum up, we have four numbers, A, B, C, and D, each of whose values can be either +1 or −1. Now here comes the algebra: consider (as mathematicians are wont to say) the number X defined as:

$$X = (A \times C) + (B \times C) + (A \times D) - (B \times D).$$

If you choose any possible combination of +1s and −1s for A, B, C, and D, you will find that X comes out either to +2 or to −2. This is a variation on the algebraic trick we had earlier; there, the answer came out to be 3 no matter what we put in; here, the answer comes out to be +2 or −2 no matter what we put in (remembering that we can use only +1 and −1 for the numbers A, B, C, and D). (See Figure 7.)

Where does this get us? We have constructed a little algebraic recipe which is guaranteed to come out to +2 or −2 no matter what results we get from the EPR experiment with the two different angles for each electron.

But wait: in any particular experiment, we measure only two of the four possible numbers, since we can perform only one spin measurement on each electron. We get a value for either A or B, corresponding to the spin measurement on the first electron, along with a value for either C or D, corresponding to the second electron. But if we perform a whole series of such experiments, with our randomizing device sending each of the electrons to each of their two magnets with equal probability, then all possible spin combinations on the electron pairs will be mea-

FIGURE 7

Whether nature obeys Bell's theorem can be established with an Einstein-Podolsky-Rosen (EPR) experiment, in which the electrons can each be sent to one or another Stern-Gerlach magnet to have their spins measured. When results from all four detectors are compiled, classical and quantum physics predict different statistical correlations.

sured, and by putting all these measurements together we can come up with an average value for our quantity X over the course of a series of experiments.

Now (and here's a second little piece of algebra) the average value of X is equal to the sum of the averages of all the bits that go into it. In the series of experiments, sometimes we will get values for A and C, sometimes for B and D, and so on. If we do the experiment often enough, all of those combinations will turn up as often as each other, so we can average together the results you obtain for each combination and thereby obtain an average value for X.

But we also know that X can only be +2 or –2. And if you take the average value of a whole series of numbers consisting of either +2 or –2, that average must without question lie somewhere between +2 and –2. (If you had the same number of +2s and –2s the average would be zero; if there are more of one than the other, the average will be tipped in one direction or the other; but there's no way the average can be more than +2 or less than –2.)

So, to come finally to the point, if we perform this version of an EPR experiment, with the spins of pairs of electrons measured in two possible directions for each spin, and calculate the average value of this quantity X, its value has to be somewhere between +2 and –2. This is what's known as Bell's theorem. It's an elementary result, following from the rules of algebra and nothing else. There's only one problem: it's not true.

Or, to be more specific, the average value of X predicted by quantum mechanics doesn't necessarily come within the range –2 to +2. Depending on just how the angles of the four Stern-Gerlach magnets are set, quantum mechanics predicts that the value of X may be as large as twice the square root of 2, which is almost 3.

This is an odd, seemingly inexplicable result. It's unarguable that the four numbers, A, B, C, and D, can have only the value +1 or –1, and getting from that to Bell's theorem is a simple matter of adding, multiplying, and averaging—not exactly controversial procedures in the realm of higher mathematics. The result seems inescapable, and yet quantum mechanics contradicts it. Bell knew perfectly well that this contradiction existed; that was precisely the point of this theorem. His insight was in realizing that this contradiction could tell you something interesting about the workings of quantum mechanics.

How can the contradiction come about? If quantum mechanics is right, does that mean that the laws of arithmetic are somehow not valid in the quantum world? This hardly seems possible, and yet so little has gone into the argument underlying Bell's theorem that it is difficult to see where we might have gone wrong.

Nevertheless, there was one part of the procedure that you might have found a little questionable: the quantity X is a combination of all the numbers A, B, C, and D, but as a practical matter any single experiment can yield only two of the four values; the other two, corresponding to the two spin directions that were not measured, are therefore unknown. To obtain Bell's theorem, we assumed that even though we didn't know the value of the two unknown spin measurements, we could still treat them in the equation defining X as if their values were either +1 or −1. Those, after all, are the only two possible outcomes of any spin measurement.

But, as Niels Bohr has been telling us for some time now, an unmeasured quantity is a meaningless thing. Cast your mind back to the original EPR proposal, in which Einstein wanted to argue that both the position and the momentum of a particle could be measured, without disturbance to the particle and with arbitrary accuracy, by making the appropriate measurement on the other particle of the pair. To Einstein, this meant that both position and momentum were real, objective quantities ("elements of physical reality," in his phrase), whose product was in principle defined to as much accuracy as you liked, contradicting the uncertainty principle. Bohr's rejoinder was that, in fact, you couldn't accurately deduce both the position and the momentum of the particle in a single experiment; you had to choose one, and forfeit knowledge of the other. This meant that in a practical sense, in terms of what you can legitimately measure, the uncertainty principle remains intact.

The difference was that Einstein was happy to combine in a straightforward if abstract way the results of two separate measurements, where Bohr insisted that what counted was only what you could consistently measure in a single, feasible experiment.

For a long time, this seemed like an airy, perhaps merely semantic disagreement. The importance of Bell's theorem is that it transforms the dispute into a quantitative one. By Einstein's logic, it's entirely permissible to combine together in the average value of X all the separate values of A, B, C, and D that have been recorded, even though they have been obtained in many different experiments. But then Bell's theorem must hold true.

But if you accept that quantum mechanics predicts an answer that doesn't obey Bell's theorem, then the way out is to follow Bohr, by accepting that you can't necessarily combine together quantities that can be measured only in separate and incompatible experiments. To state the point more forcefully, you mustn't believe that the values of the two spin measurements you haven't made can be thought of as either +1 or −1, even though you know those are the only possible values you could get were you actually to make those measurements. You can't say, in other words, that an unmeasured spin is either +1 or −1; the number is genuinely indeterminate, not just unknown. Unless you've actually measured it, it's not any number at all.

Fundamentally, this is the same old dispute that arose in 1936, when the EPR paper was published. But Bell turned this disagreement as to the nature of reality—more precisely, the nature of legitimate inferences one may make about reality—into something that could be tested experimentally.

And the answer is . . .

Before we dig any deeper into what Bell's theorem actually means, it's important to know whether experiment bears out the predictions of quantum mechanics or not. The whole point of the argument, after all, was to bring into the realm of empirical science some questions about what could and could not be inferred, from experiments of the EPR type, about the nature of underlying reality. In short, who's right, Einstein or Bohr?

Although Bell's argument opened the road to experimental tests of a new kind, it was many years before the necessary experiments could be accomplished. Alain Aspect, of the University of Paris, was the first to provide an unambiguous practical test of Bell's theorem.

Rather than pairs of electrons, Aspect created pairs of photons emitted by calcium atoms in such a way that their overall polarization state was neutral. Polarization is measured by sending photons toward a polarizing filter (like the material in sunglasses) set at some angle; a photon whose polarization is exactly parallel to the angle of the filter will go through, while one whose polarization is at right angles to the filter will be blocked entirely. Where there is some angle between the photon direction of polarization and the filter, there is some corresponding quantum mechanical probability that the photon will go through.

Although the physics is different, the logic of this is exactly the same as in a spin measurement on an electron. An electron passing through a Stern-Gerlach magnet will come out in one beam or the other; a photon reaching a polarization filter will either go through or not. The "+1" and "−1" of the spin measurement correspond to the "yes" and "no" of the polarization experiment. With that equivalence, everything we know about spin measurements holds true for polarization measurements.

Aspect's experiment was then just as we described. The two photons have no overall polarization, so that a measured polarization state in one direction for one photon necessitates an opposite polarization state for the other, just as with our familiar pair of electron spin measurements. To test Bell's theorem, Aspect arranged the paths of the two photons so that each could be sent to one of two polarization detectors, set at different angles. To ensure genuine randomization, he arranged furthermore that the choice of which detector each photon went to was not made until after the photons had left the source and were on their way; a mirror, in essence, in each photon's path could be flipped to send it to one detector or another, and the mirror-flipping was rapid enough that where the photon ended up, once it had been generated, was quite unpredictable.

In this way, the experiment tested all combinations of the detectors for each of the two photons, and obtained, over the course of time, all possible combinations of results (corresponding, in the formulation we just used, to all allowed combinations of +1 and −1 values for the quantities A, B, C, and D).

And then, of course, it was a simple matter to work out the average value of the quantity X, and see whether it obeyed Bell's theorem or the rules of quantum mechanics.

Well, not quite. We have been tacitly assuming up till now that when such an experiment is run, all results are recorded with perfect accuracy for all pairs of particles. In fact, real detectors are not so perfect. Recording the passage of a single photon or electron is no mean feat, and sometimes one will get through unnoticed. In practice, in experiments of this sort, you will obtain definite measurement records only for some fraction of the pairs produced, and since the experiment is basically a statistical one, this can potentially stymie the whole proceedings. It's as if you wanted to test the relative accuracy of two weather forecasts, but had weather records for only two days out of three. In Aspect's case, incomplete data meant that the value of X could not be determined with perfect accuracy.

But, to cut a long story short and throw a great deal of technical ingenuity aside, Aspect eventually got his experiment running with sufficient reliability and compiled a sufficient number of individual runs that he could determine the value of X with sufficient accuracy to see whether it favored the prediction of quantum mechanics, or instead obeyed the stricture of Bell's theorem.

By 1982, the answer was in, and was accepted as definitive by the physics world. Bell's theorem was not obeyed by nature, quantum mechanics came through, and our understanding of the reality of the natural world was thrown into confusion.

In which reality, once changed, can never be changed back

Bell's theorem, it's important to realize, doesn't involve quantum mechanics at all. It happened that we reached it by thinking of the numerical quantities in it as the results of some quantum mechanical measurements—electron spins or photon polarizations—but that's actually beside the point. The theorem embodies a very general view of reality, one that physicists more or less unthinkingly adhered to before quantum mechanics came along. The fact that Bell's theorem is not obeyed in the real world is telling us not so much that quantum mechanics is correct but that the old view of the world is wrong. But what exactly is that old view, and in what way or ways can it be wrong?

•

Imagine a device that creates pairs of objects and propels them separately to two different machines that measure the objects' properties in a variety of ways. Imagine also that these two machines can be adjusted in various ways, independently of each other, so that they measure a selection of different properties of whatever objects enter them. Put aside, please, as much as you can, any anticipation that the objects will turn out to be electrons or photons, or that the machines are going to be Stern-Gerlach magnets or polarization filters.

We can run this experiment repeatedly: pairs of objects are produced and sent to the machines, where results are recorded that depend on the nature of the objects entering them and on the particular adjustments that have been made to the machine, which may well vary from experiment to experiment. Perhaps at each machine there are gangs of small children twiddling knobs at random, so that the machines measure something different every time. (See Figure 8.)

Over a period of time, we compile a list of results. For each pair of particles, we note down the settings of each machine and the measurements they record. After we have compiled a long enough list, we can extract directly some probabilities characterizing the outcomes of the tests. We find that when machine 1 has certain settings and machine 2 certain other settings, then the properties of the objects are more likely to be one thing than another. Or we find that when the pairs of objects are found to have certain properties, the settings of the machine are more likely to have been this than that. Et cetera. In coming up with these correlations, we are doing nothing more than putting together a table of odds, so that as more and more results are compiled, you find you can do a better and better job of predicting what combinations of machine settings are more likely to deliver certain measured properties. There's no physics in any of this. You're just compiling results and noting, as an empirical matter, what results and what machine settings tend to go together.

FIGURE 8

Bell's theorem follows directly from the precepts of classical thinking; a version of it can be established for measurements of any properties on any pair of objects that have a common origin.

With so variable and arbitrary a setup, you might think little of substance can possibly be learned from this exercise. But let us investigate a specific scientific hypothesis: because the pairs of objects are produced by the same machine, they may have properties in common, but according to our understanding of nature, once they are on their way, nothing we do to one of the objects can possibly affect the other. Whatever happens in machine 1 can have no effect on what happens in machine 2, and vice versa. By this hypothesis we are asserting that any related or connected properties of the two objects can be correlated only by virtue of their common origin, and that no physical influence can travel from one machine to the other faster than the speed of light. These are the principles that Einstein held sacred.

What does this hypothesis tell us about the list of properties we measure at each of the machines?

It won't be surprising if we find some correlation between the outputs of the two machines, because those outputs depend in part on the objects passing through, which had a common origin. But that can be the *only* source of correlation. Our hypothesis, simple as it is, means that there is a limit to the degree of correlation that can arise between the outputs of the two machines.

Amazingly, this very general and seemingly vague hypothesis is all that's needed to obtain a version of Bell's theorem for the particular case at hand. The specific theorem we looked at in the previous section was a special case of this more general relationship.

Now, the fact that Aspect's experiment produced a result in contradiction of Bell's theorem says that nature does not play by the rules embodied in our hypothesis. Still, we have said nothing specific about quantum mechanics. This is important: regardless of anything else, Aspect's experiments (and by now others too) indicate that at least one of the assumptions going into the argument that led to Bell's theorem must be wrong.

Even if you don't like quantum mechanics, even if you think some other theory might eventually augment or supplant it, you can't go back to the old view of reality. It just doesn't work, and that is the real import of Bell's theorem.

How are we to adjust our view of reality? Before Aspect had come up with a definitive result, one possibility was of course that the real world would turn out to obey Bell's theorem, in which case quantum mechanics would have been wrong. That's not how it went. Bell himself described three other ways we might try altering our perspective so as to come to grips with our forced abandonment of the old-fashioned and now discredited picture of reality.

First, perhaps Niels Bohr was right—there is no definite underlying reality.

Second, perhaps there is some influence that can pass instantaneously from one place to another.

Third, perhaps the adjustments made separately to each detector are not really independent.

How do these ideas help us? How do they explain why Bell's theorem need not be obeyed? Let us take them in reverse order.

If the gangs of children who were twiddling the knobs on the detectors had actually got together beforehand and conspired to adjust the knobs according to some secret plan they had worked out in advance, then we could no longer maintain that the results of the measurements at each detector were truly independent, and the argument would fall down. You may respond that this is a matter under our control, and that if we take sufficient precautions we can be sure that no such conspiracy happens. If we distrust the children in our employ, we can always set up some kind of automatic roulette wheel or computerized random number generator to adjust the settings on each machine, so that there is no chance of collusion, deliberate or otherwise.

However, if we take an extreme point of view and observe that, in principle, the entire universe forms one huge but fundamentally connected quantum system (especially if we think that the universe began in a "big bang" of mysterious quantum origin), then we might argue that in fact every seemingly independent action in the universe, whether by small children or electronic random number generators, is in fact connected. In that case, whenever we do an experiment of this sort, all our preparations are subject to some sort of intimate, behind-the-scenes mutual dependence, and truly independent adjustments are impossible.

There is perhaps a vague element of science-fiction plausibility and appeal to the idea that everything is connected to everything else (and indeed mystical arguments on the holistic interconnectedness of life, the universe, and everything often take their cue from this sort of reasoning), but we must be clear about one thing: if this is the answer, then we are giving up free will. We would have to believe that when Aspect performed his experiments in the early 1980s, the seemingly random choice as to which detector each photon went to on any particular trial was not random at all, but the result of some sort of cosmic predestination that goes all the way back to the birth of the universe. The running of the experiment—not just its conception and execution, but every last detail of every particular result it came up with—was inherent in the details of the big bang itself.

This idea amounts to saving one version of classical reality by appealing to a still older version: everything is preordained (except now by quantum mechanics rather than the hand of God), and we are mere puppets following prescribed paths but too ignorant to be aware of our limitations. This does not seem to be a comfortable resolution of our difficulties.

Alternatively, if we allow instantaneous communication either between the detectors or between the objects being detected, then all bets are off and any kind of correlation between their behavior becomes, in principle, explicable. This is how Bohm's hidden variables version of quantum mechanics is supposed to work—

it explicitly incorporates a quantum potential by which the spin measurement of one electron, for example, instantaneously changes the spin of a second electron so that the two are always equal and opposite. Whether you like Bohm's theory or not, it at least brings the issue of "nonlocality" out into the open. On the other hand, as Bell once observed, Bohm's formulation resolves the EPR problem "in the way Einstein would have liked least"— because, after all, Einstein's purpose in making the EPR argument was to show that quantum mechanics was absurd precisely because it seemed to include some kind of loathsome "action-at-a-distance."

Third, if Bohr is right, it is illegitimate to speculate on the nature of physical reality underlying this sort of experiment, and as long as we have a theory that gives us the right answer, then why worry?

More precisely, the Copenhagen account of EPR and Bell's theorem requires us to say that until a measurement has actually been performed, it is wrong to think of the two objects as having a completely independent existence. They are two objects forming a single quantum system, Siamese twins intimately connected because of their common birth, and it is only when measurements are made at the two detectors that we obtain empirical recognition of them as separate particles. You, along with many others, will be forgiven if you think that this fudges the issue of whether in fact there is or is not some sort of instantaneous quantum mechanical influence pervading the whole system. Bohr might respond vigorously that this is a meaningless question (if you were to try to investigate it by adding more detectors to the setup you would be altering the whole system, and simply separating the Siamese twins at a different time), but not all of us have Bohr's sternness of mind on this score.

The possibility of simultaneity

Is it conceivable, despite all the prejudices and arguments to the contrary, that physics can admit some sort of direct influence that travels instantaneously from one part of an EPR experiment to another? Bohm's theory incorporates such a phenomenon, but by means of a quantum potential that, in the eyes of most physicists at least, is by no means an ornament to their science. Are there any other possibilities, less contrived, more appealing?

If you are a photon, traveling at the speed of light, then it's true, in a manner of speaking, that you sense no passage of time; everything becomes simultaneous. It's a standard implication of special relativity that time seems to slow down the faster you move, and light rays—or photons, since we are being quantum mechanical—move along what are called null paths, on which time has in some sense slowed to zero. You can then argue that from a photon's point of view there is no such thing as the passage of time, and that when a photon travels from point A to point B it arrives, as Omar Khayyam almost said, at the same time as off it went.

From the point of view of someone performing an EPR experiment, this is not helpful. If two spin or polarization measurements are performed at the same time, as indicated by the clock on the laboratory wall, there's no direct possibility of any instantaneous communication between them. A photon, or anything else, setting off from one place of measurement to get to the other and so communicate vital information, will get there;

by the laboratory clock, too late to influence the other measurement. To put it in a slightly more technical way, two measurements that are performed simultaneously, from the point of view of an observer in the same laboratory, cannot be connected by one of these "null paths" along which time is supposed to stand still.

So that's no good, but here's another idea. From a strictly mathematical point of view, light waves or photons that travel backwards in time are just as legitimate as those that travel forwards. From a practical point of view, however, we disregard the backwards solutions, which would correspond, for example, to a burst of light that comes at you simultaneously from all corners of the universe and converges on your flashlight. It's not that such a thing is intrinsically impossible, but that the conditions to create it cannot realistically be achieved.

But giving credence to backwards-in-time waves can lead to some curious effects, and to something called the "transactional" interpretation of quantum mechanics. The idea is to split the usual quantum mechanical wavefunctions into two parts, one traveling forwards in time and the other traveling backwards, in such a way that only the appropriate combination of these two pieces yields the correct actual wavefunction for a given interaction. To put it in more empirical terms, when two particles interact quantum mechanically, one is supposed to send out part of the wavefunction forward in time as a "feeler," provoking a response from the second particle which travels backward in time to the first particle. Feeler and response then overlap, in space as well as time, to create, all of a piece, a wavefunction linking the two particles. One is then supposed to understand from this that a wavefunction arises in apparent simultaneity everywhere, through a combination of bits and pieces of wavefunction moving both forwards and backwards in time.

And if wavefunctions are generated in this simultaneous fashion, it's no longer so difficult to understand the sort of instantaneous conspiracy between separated particles that goes on in an EPR experiment, although you have to think of one of these backwards-and-forwards feeler-and-response interchanges arising along the path that connects one measured electron with the source that created it and then interacting with another similar interchange along the path between the source and the second electron. Nevertheless, you can more or less see how to connect up the two electrons with a path that seems to involve no passage of time.

But this "transactional interpretation" is at best a sketch or cartoon of a theory. The division of the wavefunction into two parts goes well beyond the standard mathematics of quantum theory, and the means by which the "feeler" provokes the appropriate "response" is altogether mysterious. In the end it's another specious attempt to "explain" quantum mechanics by investing its inner workings with a purely speculative physical meaning; as always, you succeed only in moving the mystery from one place to another.

Not at all what Einstein wanted

In Bohm's hidden variables theory there's a quantum potential that explicitly provides a simultaneous connection between the two electrons in an EPR experiment. That's why Bell's theorem doesn't hold true. But Einstein, as Bell has remarked, would hardly have clapped his hands over this "solution," since action-at-a-distance was exactly what he objected to. In fact, Bell's theorem and the attendant arguments about how it can be avoided turn the original logic of the EPR proposal inside out.

When Einstein, Podolsky, and Rosen first formulated their argument, what seems to have been chiefly on their minds was a dislike of the probabilistic nature of quantum mechanics. If, they began, you accept that the outcomes of measurements are genuinely uncertain until a measurement is made, then a statement of the probabilities of different outcomes is the most you can hope for. But if that's so, EPR argued, then when you make measurements on two separated but correlated particles (as in the case of two electrons whose spins have to add to zero), you must also accept that some influence travels instantaneously from one measuring device to the other, in order to ensure that the pair of measurements come out right. If one electron is measured to be "up," then something has to tell the other electron to be "down," and that something has to get from the first electron to the second instantaneously.

And, continued Einstein, Podolsky, and Rosen, but especially Einstein, since instantaneous physical influences are quite beyond the pale, there has to be something wrong with standard quantum mechanics. EPR hoped that some sort of hidden variables theory would save the day. The need for faster-than-light signaling in an EPR experiment arises, so the logic went, because of the probabilistic nature of the individual measurements; but if the measurements are not probabilistic but rather predetermined, there is no need for any instantaneous signaling, because the outcomes of the measurements of the spins of the two electrons are somehow preordained from the outset. Hidden variables would provide that predetermination. In short, supply a working hidden variables theory, and you don't need the objectionable faster-than-light signaling.

It can't be emphasized strongly enough that Bell's theorem and Aspect's experiment entirely undermine this logic. As long as we were thinking only about the simplest kind of EPR experiment, in which you measure up-or-down spin for both electrons, then one measurement is fully determined by the other, and hidden variables would seem to obviate any need for instantaneous signaling. But as soon as you start thinking about spin measurements at some angle to each other, so that one measurement influences but nevertheless does not fully determine the other, probability resurfaces, and it's no longer so easy to think that the presence of hidden variables could get around the need for instantaneous signaling.

The lesson of Bell's theorem is that to understand EPR experiments, you have to accept some sort of nonlocal influence. The formulation of Bell's theorem explicitly allows the possibility of hidden variables with which the two particles are endowed at the moment they leave the machine that generates them. But if that is the only source of correlation between the results of measurements on the two particles, the theorem holds true, and since the theorem is disobeyed by Aspect's experimental results, any such version of a hidden variables theory cannot be adequate. You still need some sort of "nonlocality"—that is, some sort of instantaneous physical connection between the two widely separated particles.

In Bohm's quantum potential the nonlocality is explicit, while in the Copenhagen interpretation, it remains implicit in the statement that the two particles must be considered as a single, coherent quantum system up until the moment the first measurement is made. Far from making faster-than-light influences unnecessary, as Einstein had hoped, a hidden variables theory brings the need for such an influence out into the open.

And, as if that weren't unpleasant enough for Einstein to accept, the determinism that's concealed within Bohm's theory must in the end produce effects that look as if they are the consequence of simple probability, because experiments such as Aspect's agree entirely with the predictions of quantum mechanics. If there is a determinism at the heart of the theory, it has to keep its head down, so that you can never really be sure it's there.

What Bell's theorem has achieved here is a separation between the two issues—probability and nonlocality—that Einstein had thought were inextricably intertwined. Aspect's results prove that nonlocality is part of nature, regardless of anything to do with quantum mechanics, and as to whether the full and correct underlying theory is probabilistic or deterministic (Bohr or Bohm), the results have nothing to say. Either version works, and, after all this effort and analysis, we still have no quantitative, empirical reason for preferring one or the other.

Perhaps you're getting a little impatient. Although we've been through the ways that you can get around Bell's theorem, we still haven't really nailed down how it is that quantum mechanics gets the answer right in EPR experiments such as Aspect's. Technically, the equations of quantum mechanics predict a correlation between all the possible results (the A, B, C, and D of the four Stern-Gerlach magnets) that's larger than Bell's theorem allows, and the reason, to put it as simply as possible, is that we have to think of the paired electrons as constituting, until a measurement is made, a single, coherent quantum system, not as separate particles that happen to have some properties in common because of a common origin.

Or, to be more precise, we have to think of the electrons as constituting a single quantum system if we are interested in spin measurements, but if all we are doing is following their path through the laboratory as they leave the device that made them, then we're at liberty to think of them as separate bodies, going their own way. This is loosely analogous to the photon and the two-slit experiment, which behaves in its entirety as a single quantum system if we're interested in the interference pattern, but which breaks down into a this-slit-or-that-slit dichotomy if that's what we choose to measure instead.

Taking the Copenhagen view, we escape Bell's theorem by asserting that even to think of the electrons as separate particles with correlated properties is incorrect, so that the whole of Bell's argument about measurements at one detector not influencing measurements at the other becomes irrelevant; it's only when one measurement or the other is made that we are allowed to think of the electron spins as having any separate identity, so that trying to decide what information concerning its spin can and cannot be carried by each electron is not a physically meaningful matter.

What this means, to put it more bluntly, is that you will get nowhere worrying about what is going on with the electron spins in the time after the particles have set off from their common source but before they reach the Stern-Gerlach magnets. As always, in the world according to Bohr, you have to resist the temptation to try to infer what's "inside" a single quantum system.

Not very satisfying, is it?

ACT III

Making Measurements

Before the work of Bell and Aspect, it was possible to hold out fond hopes of one day finding a new way to understand quantum mechanics, by which its mysteries would somehow become comprehensible, and in which something like the old classical view of reality might be restored. That's no longer possible. Reality, at the most fundamental level, is not what Einstein and all his forebears had hoped it would be—had thought, indeed, that it must be. You can still believe in parallel universes, or pursue hidden variables and their concealed determinism, if that's your inclination, but you can no longer imagine that such efforts will make quantum mechanics subservient to an older, and now discredited, conception of the world.

If that lesson put an end to some long-traveled avenues of inquiry, and seemed to some a dismaying conclusion to a hopeful tale, it also encouraged thoughts in a new direction. What we mean by reality is for the most part what we perceive reality to be. If we now take to heart the idea that we are misinformed, and that reality is not what we have always thought it to be, we begin to wonder whence arises our apparently false but still insistent perception of the world.

An engineer, a physicist, and a philosopher . . .

. . . were hiking through the Scottish highlands. Coming to the top of a hill, they saw a solitary black sheep standing before them. The engineer said:

"Remarkable! Scottish sheep are black."

The physicist said:

"Strange! Some of the sheep in Scotland must be black."

The philosopher said:

"Um. At least one of the sheep in Scotland is black, on one side anyway."

Progress in science is a matter of jumping to conclusions. The trick is to jump to useful and interesting conclusions. Generalizing from small scraps of evidence may lead one astray, but sticking strictly to what limited evidence one has, and refusing to countenance anything that is not directly provable, leads nowhere at all. The scientist has to generate new ideas and hypotheses, then act upon them.

Perhaps this is why the Copenhagen interpretation seems so meager. It seems to deny the essential right to speculate beyond the evidence at hand, and to deny, therefore, the scientist's ability to make progress. Hence, similarly, arises the appeal of the Bohm approach, which seems to restore the idea of a genuine objective world beyond (perhaps "behind" is a better word) mere observations—a world whose structure scientists can speculate about and explore further. In Copenhagen, the observations are all you get; with Bohm, there is the chance of digging deeper.

On the other hand, revolutions in science have sometimes come about because conclusions that everyone had jumped to, because they seemed so obvious as to be self-evident, turned out to be questionable. The transition from Newtonian to Einsteinian dynamics forced some fundamental reexamination of "obvious" conclusions: time is not the same for everyone; the speed of light cannot be surpassed. Does the Bohm versus Bohr debate amount to such a revolution in the fundamentals of physics?

The essence of many criticisms of the Copenhagen interpretation is that it denies something scientists call "objectivity" in their descriptions of the world, and thereby denies the essence of the scientific picture of the world. A belief in objectivity is generally taken to mean a belief in the idea that the world exists independently of us and our measurements of it, and that as we build up a coherent scientific account of the world around us, we are constructing an increasingly detailed and precise image of something that already exists. We are gradually seeing through a fog to build up a picture of an elephant that was already there. By suggesting that the appearance of the world depends on what observations we make, and on how we choose to experiment upon it, the Copenhagen view explicitly rejects the idea of a preexisting world whose attributes we are trying to tease out. And that seems to make the world to some extent a matter of our whim and fancy.

The mathematician and relativist Roger Penrose, for example, says, "Like Einstein and his hidden-variable followers, I believe strongly it is the purpose of physics to provide an *objective* description of reality"—a description, in other words, that does not depend on the unpredictable choices this or that experimenter might make. Take away this objectivity, as the Copenhagen view does, and you turn the foundations of science to sand.

But let us dissect more carefully this idea of objectivity. Penrose uses the phrase "description of reality," and followers of

Bohr rather than Einstein will insist that the Copenhagen interpretation provides exactly that—a "description" of reality that
is objective in the sense that everyone armed with a quantum
mechanical education can do the same experiments and come
to the same general conclusions. What does quantum mechanics provide, if not a means of analyzing any experimental setup
you care to imagine, and predicting from it the range of possible results you will encounter? Is this not an objective procedure? There is the fact that the predictions are probabilistic in
nature, but that is not to say they are unobjective—everyone
will agree on what the relevant probabilities are; a series of
experiments done by the same or different people will produce
a list of results in accord with those probabilities; and the outcomes of individual experiments will invariably be among the
range of possible outcomes that quantum mechanics allows.
Penrose himself goes on to say that it is not the indeterminism
of quantum mechanics that bothers him. So what exactly is the
trouble here?

The answer is that by "objectivity" scientists traditionally
mean something more than an ability to agree on the results (or
range of possible results) of experiments. They mean, more profoundly, that the experimental results they obtain refer to an
objective reality, objective in the sense that *all* experimental
results are consistent with the *same* underlying reality. It is this
second idea that Copenhagen will not allow: as we have seen
with EPR experiments in particular, one must scrupulously
avoid the temptation to suppose that the results of two incompatible experiments (two experiments that physically cannot be
done at the same time, on the same system) must yield harmonious results.

We can divide the scientists' notion of objectivity into two
parts. According to something we might call "weak objectivity," it's essential to the function of science that all scientists can
agree on the rules, and can unambiguously agree, when faced
by a specified experimental setup, on what can happen and

what can't happen. This indeed seems essential to science, and quantum mechanics, in whatever interpretation we choose, adheres to this principle.

But what we can call "strong objectivity" goes further, and declares that the picture of the world yielded by the sum total of all experimental results on all possible pieces of the world is in fact not just a picture, but really is identical to the objective world, something that exists outside of us, and prior to any conception or measurement we might have of it.

Going from weak to strong objectivity—which amounts to going from a belief in the reality of your experimental results to a belief in the reality of a partly perceived world which those results represent—is a matter of jumping to a conclusion without any real evidence. Just the kind of jump, in fact, that has allowed science to advance. And in classical physics, you run into no trouble making this jump. It's fundamental to the classical view of the world that measurements made by scientists are measurements of a genuine underlying reality.

Nevertheless, the jump is not demanded by classical physics. It's consistent with the way classical physics works: scientists are at liberty to assume that all their measurements refer to an objective underlying reality without running into any kind of contradiction or difficulty, but it's not an assumption that science itself forces upon the scientist. In classical physics, in other words, weak and strong objectivity are in practice quite indistinguishable, and there is nothing to say a scientist should not hold to the strong objectivity line. But at the same time there is no compulsion to go from one to the other. Weak objectivity is a minimum standard that scientists must accept in order to pursue their work: they must be able to agree on rules and procedures, to make unambiguous comparisons of experimental results, to have assurance that the same experiment done by different people at different places and times will always yield consistent results. There's an almost irresistible temptation, bolstered by hundreds of years of successful scientific investigation, to make the leap from weak to strong objectivity. But it remains a leap of faith, not a scientific necessity.

And the new ingredient that quantum mechanics brings to this discussion is precisely that it drives a wedge between weak and strong objectivity. You can still agree on experimental definition, procedures, and results, says the Copenhagen interpretation, but you can't agree on the existence of a harmonious underlying reality. The EPR experiments, Bell's theorem, Aspect's results—all these evidences amount to a demonstration that disparate fragments of information, obtained from different experiments, can't always be assembled into a single portrait of the physical world. In classical physics, the surface temperature of Pluto is a unknown number that has independent and objective existence, and observations can yield successively better estimates of that number. In quantum mechanics, the spin of an electron is genuinely indeterminate—not the same thing as "unknown"—until you measure it, and a person who measures the spin of one electron of an EPR pair in an up-down magnet has no way of comparing his result with that of a person who measures the spin of the other electron in a left-right magnet and arriving at a mutually agreeable account of what the spins of the pair of electrons "really" were. Strong objectivity cannot be sustained.

As with any revolution, two responses are possible: you can accept it or you can fight it. Accepting it means accepting that belief in an objective reality cannot be maintained, and that we will have to make do with what I have called weak objectivity. Fighting it means, like Bohm, looking for a way of restoring to quantum mechanics an underlying reality, even if it remains an unseen and undetectable one. In this sense, the traditional Copenhagen view is, despite its hoariness, the revolutionary philosophy, while Bohm and his followers (even Einstein, it must be said) are the counterrevolutionaries, seeking to turn back the clock to a simpler time.

We have seen that Bell's theorem, for all the illumination it provides, still cannot tell us that one theory is right and the other wrong. Because Bohm's theory is mathematically identical to standard quantum theory, it's possible that no experiment can ever really decide the issue. It's then a matter of philosophical taste which view one prefers. Practically speaking, most physicists implicitly rely on the Copenhagen view, because it's the simplest; Bohm's version of the subject adds mathematical complication without generating any new results. Even so, many of the physicists who implicitly follow the Copenhagen line start to squirm a little when faced directly with its deepest implications. In time, perhaps, as the experimental results of the last few years sink in and scientists accept them at face value, the revolution will be completed. And as for hoping that one day some ingenious physicist might come up with a theory that's "deeper" than quantum mechanics, and newly illuminates our understanding of reality, it's possible, of course. But there seems to be no practical need for such a theory, and perhaps in any case such a theory is not desirable. Would a theory that fully restored determinism to the world put us back in the eighteenth century again, with a view of the world in which everything's strictly cause-and-effect, nothing happens by chance, and, therefore, all our thoughts and actions are predestined? Perhaps the ineffability of reality, according to the Copenhagen view, has its advantages. . . .

In the meantime it's important to understand that scientists still seem to be able to do science in more or less the traditional manner, without worrying about objectivity too much. Perhaps, over the course of centuries, scientists have got so used to the idea that something like strong objectivity is the foundation of their knowledge that they have come to believe it is an essential part of science; that without this most solid kind of objectivity, science would become pointless and arbitrary. But the success of quantum mechanics tells us that this simply isn't so. Even if you don't like the Copenhagen interpretation, which denies that there is any such thing as a true and unambiguous reality at the bottom of everything, you have to admit that it works. And the fact that it works—meaning simply that no one has come up with an experiment that has proved or even could prove it to be false—shows that science can go on despite a loss of this strong kind of objectivity. This is a statement not of philosophical preference but of empirical observation: quantum mechanics works, the rest of science still works, even without an old-fashioned belief in objectivity.

But that brings us back to a question we put aside some time ago. If it's true that "reality" is not the solid, indisputable thing we used to think it was, then why does it seem, for all intents and purposes, to behave as if it were? We need to worry about interpretations of reality when we think about EPR experiments, but not when we send space probes into the outer solar system. The planets, at least, seem to act as if they really are solid, incontrovertible spheres. How come?

The one true paradox

The Einstein-Podolsky-Rosen experiment; the two-slit apparatus in which a photon seems to follow two paths at the same time; delayed-choice experiments, in which a particle seems to need to know in advance what is expected of it, so that it can appropriately gear itself up for a measurement that occurs later: you will sometimes find all these things referred to as "paradoxes." It's clear by now, I hope, that none of them is really a paradox. They're weird and bewildering, certainly, and force a fundamental change in the way we think about reality, but there's nothing strictly paradoxical in them; what they contradict is not themselves (the true definition of a paradox) but our prior expectations of the way we think things ought to be. And that's just too bad. It's up to us to make sense of nature; it's not nature's obligation to behave as we would like.

So let us simply accept all that we have learned so far. And since none of the other "interpretations" of quantum mechanics we have looked at has brought us any real peace of mind—because, of necessity, they simply push the weirdness around, from one place to another, but cannot make it go away—let us agree to stick as far as possible with the Copenhagen interpretation, which has the virtues of simplicity and necessity. It takes quantum mechanics seriously, takes its weird aspects at face value, and provides an economical, austere, perhaps even anti-septic, account of them.

But it has one flaw, a genuine paradox, which we must now finally tackle. It asserts that measurements can be made, but does not explain how they can be made. A measuring device, according to the Copenhagen view, is a machine that converts an inde-terminate "some of both but neither one thing nor the other" quantum state into a definite "either one thing or the other" clas-sical state. The "half-up, half-down" electron becomes definitely either up or down. Measurement, therefore, is the step we must take to get from the fuzzy quantum world to the sharply focused classical world. It's a necessary step, and a feasible one, evidently, since experiments can indeed be done and the world around us indeed seems to exist, but nowhere in Copenhagen can we find an explanation of how measurement comes about.

Why is this a paradox rather than simply an omission? Because we would like to believe that quantum mechanics is a fundamental theory of physics, and so would like to believe that the behavior of all objects—gloves, billiard balls, cats, the Moon and the Sun—can in principle be attributed to the prop-erties of their components, which are the atoms and electrons and photons that quantum mechanics describes. And among the large-scale objects we would like to explain in this way—or at least like to think we could in principle explain—we must include measuring devices such as photon detectors and Stern-Gerlach magnets.

According to Copenhagen, quantum systems exist in indeterminate states until they are measured by measuring devices that have only definite states; the electron's spin is unambiguously measured to be either up or down; the photon is either detected or it is not. In measuring devices there are no indeterminate states, and yet measuring devices are, if we look at them in the ultimate detail, built from quantum objects. We have arrived at a perspective according to which elementary quantum systems such as electrons and photons can exist in indeterminate states, but complicated quantum systems, at least the ones we use as measuring devices, exist only in definite states. It's as if we are saying, in some poorly articulated way, that quantum mechanics is a fundamental theory, and yet somehow doesn't apply to the special kind of devices we use to make measurements.

One way out of this paradox is to declare that quantum mechanics is not after all a universal theory, and that somehow it must be modified or even overturned when we use it to describe large-scale or macroscopic objects such as measuring devices. But this is a last resort, and Niels Bohr himself certainly didn't want to follow this path.

But Bohr "solved" the measurement problem only by fiat: he just declared that (as every physicist knows) measurements can indeed be made, and didn't concern himself any further with how. In practice, the Copenhagen philosophy doesn't cause any trouble. But intellectually, Bohr put a barrier of logic between the quantum world and the apparently classical macroscopic world, preempting any attempt to explain how the barrier arises.

The measurement problem, this one true paradox of the Copenhagen interpretation, is in fact what Schrödinger's cat is trying to tell us. The cat's survival or demise, you remember, is linked to a quantum measurement whose two possible outcomes each have a fifty-fifty chance. And if the quantum system is in an uncertain state, encompassing both possibilities but not yet settled in either, how is that quantum state transformed into an all-too-definitive dichotomy, in which the cat ends up either alive or dead?

We need to define exactly what this problem is about. To avoid grisliness (though we will get back to it) let us put our cat in a safe place and think about measuring the spin of an electron with a Stern-Gerlach magnet. Beyond the magnet we need some sort of electron detector, so that we can see which path the electron followed, up or down, and this detector can be electronically or mechanically connected to a big pointer on the laboratory wall that points UP when the electron comes out along the up beam and DOWN when the electron comes out along the down beam.

Now, if we have an electron that we know to be in an up state (because we have just extracted it from the up beam of a prior Stern-Gerlach magnet), we know that it must trigger an UP response from our pointer. Similarly, an electron known to be down will trigger a DOWN response. That's simple and definitive enough, but not at all informative; if we already know the spin state of the electron we don't learn anything new from another spin measurement with the same orientation.

Suppose now that we have extracted an electron from the left-hand beam of a horizontal Stern-Gerlach magnet; it's in a definite left state, of course, which means it has equal probability of coming out up or down in a vertical spin measurement. We describe it by saying that its wavefunction with respect to an up-down measurement is "half-up, half-down." What happens when this electron goes through our Stern-Gerlach magnet?

An up electron produces an UP response from the pointer; down leads to DOWN. So, inevitably, "half-up, half-down" leads to "half-UP, half-DOWN." And that's the problem: as far as we know, the big pointer on our measuring device can only be either UP or DOWN, one or the other, for sure, but not both.

And yet it should apparently respond to an indefinite electron wavefunction by getting stalled in an indefinite state itself. Just as for the electron itself, a pointer state of half-UP, half-DOWN doesn't mean that it actually is one thing or the other—it's that familiar "neither one nor the other but partly both" state that we have used to represent indefinite quantum states.

And we know, of course, that real pointers on real detectors in real laboratories are always either UP or DOWN. We don't even know what it means to say that it could be in a half-UP, half-DOWN state. We only know that such things are never seen.

In short, quantum mechanics alone doesn't seem to allow measurements actually to take place. If we try to describe our detector using quantum language—which, if quantum mechanics really is fundamental and universal, we ought to be able to do—then indefinite states generate indefinite measurements.

This is the measurement problem recognized long ago by Bohr, whose high-handed but effective prescription was simply to declare that a measurement must happen. In effect, Bohr implied, without ever explaining how it came about, that macroscopic classical devices such as pointers on detectors could not fall into uncertain states. They had to be one thing or the other. We just don't allow, in other words, pointers to occupy states such as half-UP, half-DOWN, but no explanation for this law was given. Certainly, it seemed, quantum mechanics itself supplied no such rule of impossibility.

Measurement, in the Copenhagen interpretation, is therefore a magical, unexplained happening, if not an act of God then certainly an act of Niels Bohr.

How can this dilemma be escaped? There are only two possibilities: either quantum mechanics is incomplete, and some other theory must be adjoined to it to explain how measurements happen, or else quantum mechanics contains some recipe within it, unelucidated up until now, that can in some way transform an indefinite into a definite state. Before we think about changing quantum theory, we should think hard about the second of these two possibilities. Following that thought, we would like to conclude that in some way all the fundamental strangenesses of the quantum mechanical world do not, when we are thinking about the macroscopic and tangible world around us, have any practical or detectable significance. Which brings us at last to the most puzzling question of all: where does the weirdness go?

At a loss for words

The electron, before it's measured, is in a state that we have called "half-up, half-down," or more verbosely "neither up nor down but partly both." Once it's measured, it's definitely either "up" or "down," and that's a statement we find altogether more comprehensible. It's like saying that the coin on the floor that we haven't yet looked at must be either heads or tails. Technically, the first kind of uncertainty, pertaining to the indefinite quantum state before a measurement takes place, is called a "superposition"—to indicate, loosely, that both possibilities are simultaneously and equally present. The second kind of uncertainty, where we know something definite has happened but we just don't know what, is sometimes called a mixture, for reasons we'll get to in a moment. Superpositions are quantum states, mixtures are classical states, and the measurement problem, now that we're equipped with the right language, is a matter of getting from one to the other. How does a superposition turn into a mixture?

Although we've gradually been getting comfortable with the idea of superpositions, exemplified by the half-up, half-down electron, you may still be confused about what exactly such a state is, particularly when we got to literally unimaginable superpositions such as the half-UP, half-DOWN pointer and the half-dead, half-alive cat. It's not so much that we really know what we mean when we say an electron is both up and down but at the same time neither really up nor down, it's just that we've grown a little familiar with the idea, and since we don't have much of an idea about what an electron looks like in the first place, it's perhaps a little easier to accept without protest that it can inhabit a weird and indescribable state. But when we try to translate any half-formed idea of what we mean by a superposition into the world of cats and pointers—then we run into a wall. The same language that we used about electrons, however little it meant then, means nothing at all now.

You should take heart first of all from the fact that physicists do not have any secret key to understanding these things, or some peculiar faculty of brain that lets them grasp concepts that the rest of us find confusing and unenlightening. They have to use the same words that the rest of us use, but they've had time to grow accustomed to odd ideas, clumsily expressed. There's a genuine difficulty of language here, in that we're trying to use words and concepts that apply to the world around us—the world of measured properties and actual results—to the hidden quantum world, in which properties are undetermined and results are as yet mere possibilities.

A classical mixture is an unfamiliar term for a familiar idea. It describes a system that is in some definite state out of a range of possible states, except that we don't know precisely which one it happens to be in. A tossed coin, before you look at it, is either heads or tails, but you don't know which. The pointer on an electron spin detector is, after the electron's passage, either up or down, but you don't know which until you look. A cat in a box is, we think, either dead or alive, but before you open the box you cannot say which. A classical mixture defines our ignorance: the system is beyond doubt in one state or another, but because we don't yet know which, we must be content with a list of probabilities that it is in fact in this state or that state or the other. For a somewhat more complicated example, think of tossing two dice. There are thirty-six possible outcomes—six different numbers on one die, each of which can be coupled to one of six different numbers on the second—corresponding to eleven different totals from two (a double one) to twelve (a double six). These two are the least likely outcomes, each with only a one in thirty-six chance of occurring, while the most likely sum is seven, which can occur in six different ways, and therefore has a probability of six chances in thirty-six, or one in six. The appropriate classical mixture to describe this situation is a set of eleven probabilities for each of the eleven possible two-dice sums.

The term "mixture," incidentally, is somewhat inappropriate when we are thinking about single systems. Originally, the terminology arose in the early days of statistical mechanics, when physicists were working out how to deal with huge systems of atoms constituting gases, and trying to understand the gross properties of volumes of gas in terms of the motions of all the constituent atoms. In this case, you really were dealing with a mixture: you couldn't say what any particular atom was doing, but you could calculate on average how many atoms were moving at this speed, how many at that, and so on. The system formed a genuine mixture of atoms, in other words, with each atom representing one particular possibility out of all the things the atoms could be doing, and with each possibility typified somewhere in the gas by some number of atoms. Applying the term "mixture" to a single instance—a single roll of the dice—should not be taken to mean that all the possibilities somehow coexist in one example. The mixture refers to the range of possibilities that this one example may in fact represent.

A quantum superposition, by contrast, is quite a different thing. It refers to an object that's in no definite state, because no measurement has yet been made. In a superposition, unlike a mixture, we cannot say that the object is actually in one state or the other, which we happen not to know; instead, the superposition contains all possible outcomes but is equivalent to none of them.

Words begin to seem slippery, and meaning elusive, as we try to define such things. A quantum superposition corresponds to nothing we have encountered in everyday life, and therefore baffles us. The best recourse is to look at what quantum superpositions mean in practice, to understand them by their implications.

The problem then, of course, is that real electrons and photons—that is to say, any electron or photon that has been detected or measured—are necessarily in some definite state. Once we make a measurement, the electron's spin really is either up or down, the photon really is either here or there. This is familiar, comfortable language, and inevitably we try to describe strange and unfamiliar things in more or less familiar ways in order to bring the strangeness within our compass of understanding. Our familiar language is appropriate only to quantum objects that have been measured, and are therefore no longer in superposed states, and yet we wish to use some version of the same language to describe the prior, unmeasured state, for which by definition it is unsuited.

In other words, the measured properties of objects are in some sense secondary to or derivative of the unmeasured state we wish to understand, so that we end up trying to describe the primary structure of the theory—the quantum superposition—in terms of language that becomes appropriate only when a measurement has been made and the superposition is lost.

This is more than mere semantics. When we describe an unmeasured spin as being both up and down, but yet neither up nor down, we are implicitly assuming that the prior state, the unmeasured electron, is somehow composed of the possibilities that emerge later, after the measurement is made. This is unwarranted and misleading. Suppose that we were all set to make an up-down measurement of an electron's spin, and were struggling with words to describe the "both up and down, neither up nor down" of the electron's disposition. But now suppose someone sneaks into the lab and rotates our measuring device by ninety degrees, so that what would have been an up-down measurement suddenly becomes a left-right measurement. Suddenly, we have to abandon the up-down language and replace it with similar statements using the words "left" and "right" instead. Nothing has happened to the electron; a "half-left, half-right" state is the same as a "half-up, half-down" state; and we shouldn't suppose that the change of language corresponds to any change in the electron's state. But then, as a corollary, we mustn't mislead ourselves into thinking that a half-up, half-down electron actually contains physically identifiable or distinguishable up and down bits, any more than the same half-left, half-right electron actually contains physically distinguishable left and right bits.

Words fail us. But if we cannot, by tinkering with niceties of language, hope to arrive at a verbal definition of a "superposition" that is clear and unambiguous, we can only hope to become familiar with superpositions by looking at how they behave. Ask not what a superposition is; ask rather what it can do for you.

Can a quantum superposition be seen?

The plain answer is no, of course: by definition, quantum superpositions are what exist prior to measurement. As soon as a measurement is made, the rules dictate that one particular result out of the range of possible results must be obtained, so that the simultaneous presence of two different states can never be directly demonstrated. As always, the mysterious, inscrutable parts of quantum mechanics remain hidden from our eyes (which is why they are mysterious and inscrutable: if quantum superpositions were to be found in the world of our senses, we would long ago have developed a language for them, and would not find them unfamiliar).

We can, nevertheless, see the effects or consequences of quantum superpositions. The two-slit experiment remains the prime example: a single photon somehow passing through a two-slit apparatus creates an interference pattern in a way that a strictly locatable classical particle could not possibly do. If you insist on saying, in particle language, that the photon must go through one slit or the other, you can't understand the interference pattern, which in wave terms requires that some contribution comes from each slit. It's only by accepting the idea of the superposition of states, in which there are coexisting pieces of wavefunction describing photons that go through both slits, that you can begin to understand how interference in such a case can happen. But at the same time you have to be wary of imagining that the parts of the wavefunction associated with each slit have anything to do with actual photons going through actual slits—things get tricky, as we have seen, as soon as you start imputing a traditional degree of reality to these equipresent bits and pieces of wavefunction.

These familiar examples of superposition—the two-slit experiment, electrons prior to spin measurement, or photons prior to a polarization measurement—apply to single objects, individual quantum creatures. Can large systems—ones composed of lots of photons, electron, or atoms—behave in the same way? Not easily, it turns out, but there's at least one reputable and instructive example.

The example comes from certain kinds of superconducting devices, which when cooled to suitably low temperatures conduct electricity without any resistance. An electric current set up in a ring of superconducting material will flow around and around forever, without loss. Because of resistance, an electric current set up in a ring of copper wire, for example, will come to a halt almost the instant the motive power is turned off.

This isn't the place to discuss in any detail the complex and demanding theory of superconductivity, which wasn't put together until almost half a century after the phenomenon itself was first discovered, back in 1911. All we need to know here is that the electrons in a superconductor move in a coherent fashion. In both an ordinary copper wire and a superconductor, electric current amounts to a collective flow of electrons, but in a copper wire, the electrons jostle about like a huge crowd trying to follow a parade down narrow streets. They bump up against the copper atoms of the wire and each other, and the incessant jostling and obstructing and diversion amount to a resistance to the flow. It takes effort—in the form of a voltage applied to the wire—to keep all the electrons moving along.

But in a superconductor, the electrons abandon all sense of individual identity and purpose, and move as one. It's not that every electron is doing the same thing as every other electron, but that they all move in a coherent fashion. A human example of such coherence is the wave that crowds perform at a baseball game. If everyone in the stadium stands up and sits down at random, all you'll see is confusion, but when everyone harmonizes their motion, a wave appears to travel around the stands. An interesting phenomenon occurs here: as one person, at one place in the crowd, is standing up, another person, perhaps on the opposite side of the stadium, is sitting down. As the wave moves around, you could spot any number of such pairs of people whose up-and-down motions are exactly harmonized. The electrons in a superconductor behave something like this. They link up in pairs, not in the sense of being physically linked, but in the sense that their motions are coupled even though the electrons are a long way apart.

Quantum mechanically, the upshot of this harmony is that all the electron pairs in a superconductor are described by the same wavefunction (all the pairs of people in the baseball stadium are doing the same thing, with systematically arranged timing). When a superconductor is joined end to end to make a superconducting ring, something else happens: because all the electron pairs follow a single wavefunction, that wavefunction has to wrap around on itself in the ring without any abrupt change. Among the implications of this wavefunction continuity is that any magnetic field passing through the interior loop, connected as it is to the current flowing around the loop, becomes "quantized," meaning that it can take on only certain discrete values.

It's not too difficult, these days, to make a superconducting ring centimeters or more across, and yet even this large an object should, if quantum theory is correct, be correctly described as a single quantum system characterized by the current flowing through it and the magnetic field trapped inside it. Even though it's a large, compound object, with an electric current and magnetic field that arise from the motion of countless electrons, it is, or should be, a single quantum system, just as an individual electron or photon is a single quantum object.

An unbroken ring, once set up, will carry current forever. It remains, in other words, in a fixed quantum state. A more interesting state of affairs can be created by putting a small gap somewhere in the ring, so that the electrical circuit is not quite a full circle but has a slightly resistive "weak link" somewhere along its circumference. As long as the gap is narrow, the superconducting current will continue to flow, being able in effect to jump across the gap. But the presence of the gap allows, roughly speaking, the magnetic field threading the ring to come and go a little; it is no longer strictly confined, and additional increments of magnetic field can jump into the ring, through the gap, or out of it the same way. Now we have a quantum system, characterized by a certain intensity of magnetic field, that can jump from one state to another.

Difficult and elaborate experiments have been performed in the last few years in which the imposition of an external magnetic field is used to control the stability of the individual quantum states of the ring, and to influence the ease with which it can jump from one state to another. The system does indeed jump in the manner prescribed by quantum mechanics, and certain more sophisticated results can be obtained which indicate that the magnetic field can exist in a genuinely superposed state connoting the simultaneous presence (to revert to our loose use of language) of different magnetic states of the ring.

Of course, these superpositions, as we said at the outset, cannot be directly demonstrated, because any measurement of the ring's magnetic state forces it to adopt one specific value or another. Rather, certain measurements of other properties of the ring, which don't interfere with its magnetic state, are consistent with the existence of a superposition of magnetic states within the ring.

In short, everything works the way quantum mechanics says it should work. What does this tell us? Mainly, that in this one example at least it is possible to have a macroscopic system that lives by the rules of quantum mechanics. That's an important conclusion, because it had been thought from time to time that a simple way of escaping the measurement problem would be to say that macroscopic systems, as opposed to individual quantum objects, would not obey the elementary rules of quantum mechanics, and so could not sustain superposed states. The superconducting ring is a counterexample to that hypothesis: it's definitely a macroscopic system, but it evidently behaves just as unadulterated quantum theory predicts. You can't get around the measurement problem by supposing that for some reason big objects and systems obey a different set of rules.

You might say that the traditional two-slit experiment is a macroscopic demonstration of the correctness of quantum mechanics, since one ends up with a large apparatus and a result—the appearance of an interference pattern—that's easily visible to the naked eye. But in a two-slit experiment, the fundamental quantum object is the individual photon going through the apparatus, whereas in the superconducting ring, the quantum states represent collective motions of trillions of individual electron pairs acting together.

In short, it's not the case that just putting together a lot of individual quantum objects necessarily erases quantum behavior. On the other hand, a superconductor is an exquisitely special and unusual creature, and electrons in large systems generally don't get along so well together. If it's true that superconductors exhibit large-scale quantum behavior only because of the very special conditions under which they exist, then perhaps we should ask in what way ordinary, nonsuperconducting systems are different, and if that difference has anything to do with why, for the most part, they do not apparently show quantum behavior.

Like peas in a box

A superconductor can behave like a large-scale quantum object, but most macroscopic systems—pieces of ordinary copper wire, pointers on detectors, cats—do not. The distinguishing feature of a superconductor is the orderliness or coherence of the motion of its electrons, which stands in contrast to the disorderly motion of electrons in a copper wire and, in the same vein, the random jiggling around of atoms in most large objects. Might this distinction between orderliness and disorderliness have something to do with the quantum mechanical behavior of superconductors and the apparently non-quantum mechanical behavior of most everything else?

Here's a novel way to entertain your friends one evening, if gin rummy no longer thrills you. Make a transparent Lucite box, a foot or so on a side, and seal a dozen Ping-Pong balls inside it. Now shake the box around vigorously, so that the Ping-Pong balls rattle around this way and that, and enlist one of your friends to make a video recording of the event. Then put the tape in your VCR and watch the recording.

Well, maybe it wouldn't be all that entertaining. But here's an interesting question: could you tell if the recording was being played forwards or backwards? In fact, there's no way to tell. The detailed motion of the Ping-Pong balls in the box might well look different, depending on which way the tape is being played, but there's nothing in those small differences that can reveal whether time is running forwards or backwards.

This illustrates a basic principle of physics: the laws of mechanics governing the motion and collision of Ping-Pong balls in a box look the same whether time is running forwards or backwards. Two Ping-Pong balls head towards each other, collide, and move off in opposite directions, obeying the relevant rules of conservation of energy and momentum. But the same collision run in reverse also obeys the same laws of physics, and therefore constitutes a physically allowable event. In fact, if you could stop the Ping-Pong balls as they emerged from the collision, exactly reverse their motions so that they headed back towards each other, then they would collide and come out exactly on the same tracks along which they originally went in, and with the same speed. The two collisions are precise mirror images in time, so to speak.

And, because of this, a dozen Ping-Pong balls rattling around inside a box execute a complicated interacting series of collisions which, for all their complexity, are exactly reversible in time. Playing the movie backwards gives a set of collisions that's exactly consistent with the laws of physics, so that unless you are told there's no way of knowing which way time was running when the movie was actually recorded.

Here's a slightly different thing to think about: suppose one of the Ping-Pong balls is red, but the other eleven are white. Does this make any difference? In fact, no: you could track the path of the red ball as it moved around among the others, this way and that, but you still couldn't tell whether the movie was running forwards or backwards.

Another variation: instead of Ping-Pong balls, let's use dried peas instead. Put a thousand green peas into the box, and then add fifty yellow peas, all in one corner. Now start shaking the box around.

An interesting thing happens: at first, you see all the yellow peas in one corner, but as the box is shaken around they start to mix up with the green peas. Once all the peas are thoroughly mixed, we're back as we were before—there's no way of telling if a movie of the peas is running forwards or backwards. But for the first few moments that's not the case; right at the beginning, you can see that the yellow peas start out in one corner, and get mixed up, but if you run that section of the movie in reverse, you would see the yellow peas separating themselves out and ganging up in the corner. That kind of thing doesn't happen, you think, and you decide the movie must be running backwards.

What's going on here? The laws of physics aren't changing, and every individual collision between peas or Ping-Pong balls is, as it must be, reversible in time. But somehow, if the number of things bouncing around is large enough and if some distinct fraction of them starts out in some unusual configuration, there's at least a fleeting ability to tell which way time is running. But that distinction is lost again once the peas are thoroughly mixed. (See Figure 9.)

So here's a conundrum. The motion of a thousand thoroughly mixed green and yellow peas can be watched ad infinitum—literally—without giving the observer any clue which way time is really flowing. But when the yellow peas are first being mixed in, you see what physicists call an irreversible process: the yellow peas get mixed into the green ones, and the mixing looks perfectly fine if the movie is running forward and quite incorrect when the movie is run backwards. If all the fundamental laws of physics in this case are indifferent as to the direction of time—reversible, as this property is called—then where does this evident irreversibility come from?

More to the point, you might say, what has shaking peas around in a box got to do with the mysterious act of quantum

FIGURE 9

This happens . . .

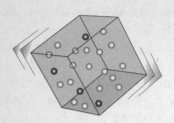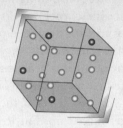

This happens . . .

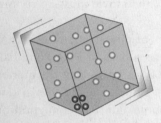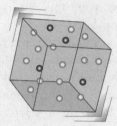

This hardly ever happens . . .

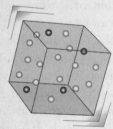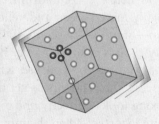

Although the laws of mechanics look the same whether time flows forwards or backwards, and although shaking peas around in a box involves those laws and no others, going from an unmixed to a mixed state seems to define unambiguously a direction of time.

measurement? The connection, which so far is no more than a hint, is this: going from a simple situation involving a few particles to a more complex situation involving many particles, and distinct groups of particles, seems to introduce a change from reversibility (you can't tell which way the clock is running) to irreversibility. We haven't yet reached the point where we can define what a quantum measurement amounts to, but irreversibility certainly seems to be involved: once an electron's spin has been established to be "up," the prior uncertain state is lost; once Schrödinger's cat has been pronounced dead, a live cat cannot be conjured from it—still less an uncertain half-dead, half-alive cat.

But shaking peas around in a box has nothing directly to do with quantum mechanics. The whole process can be understood entirely and correctly in terms strictly of classical physics. Before tackling the irreversibility of quantum measurements, therefore, we need to understand more clearly the irreversibility that can arise in classical physics alone. The connection turns out to be more intriguing and enlightening than you might suppose.

More than you really wanted to know about dried peas

We have a movie of our plastic box filled with peas, with the fifty yellow peas starting out in one corner and being mixed into the thousand green peas. At first, as the yellow peas get mixed into the rest, there's something obviously irreversible going on, but once the mixing is complete, reversibility seems to be restored, and we can no longer tell which way time is running—whether the movie is going forwards or backwards.

Here are three more perplexing thoughts. First, the movie, when it's run in reverse, portrays a series of collisions and motions that obey the laws of physics every bit as much as the same movie running forwards. Any collision between two Ping-Pong balls or two peas transforms, when time is reversed, into an equally possible collision. If you were to look at individual collisions in the movie, even at first when the yellow peas are being mixed in, you still couldn't tell which way time is running. It's something about the collective behavior of all the peas that betrays the direction of time, and yet that collective behavior amounts fundamentally to nothing more than a collection of individual collisions, which run according to time-reversible laws. How does a collection of collisions, which by themselves are indifferent to the direction of time, produce as a whole an obvious direction of time, in the form of yellow peas getting mixed into the green ones but not getting unmixed?

Second, here's a theorem due to the great turn-of-the-century mathematician Henri Poincaré: if you regard every possible configuration of yellow peas and green peas in the box as an allowable "state" of the whole system, and if you shake the box endlessly at random so that it constantly moves from one such state to another, then in the fullness of time (meaning, to be precise, if an infinite amount of time is available) the system as a whole will visit every possible allowed state. In particular, since the starting configuration, with all the yellow peas in one corner, is an allowed state of the system, you will definitely get it back again if you keep shaking the box long enough. A long enough movie will show the initial "unmixed" state arising again, purely by chance. In which case, you would in fact not be able to say definitively whether a movie segment showing all the yellow peas coming briefly together was running forwards or backwards in time. According to Poincaré's theorem, both mixing and unmixing are possible, and anything that's possible must eventually happen.

Third, if the movie run in reverse portrays motions and collisions that obey the laws of physics just as well as the movie run forward, what is to stop you from measuring, with great

care, the precise physical action of the shaking and then applying those motions in reverse to re-create the initially unmixed
state by design, rather than simply waiting around for nature
and Poincaré's theorem to work their inevitable effect? (You
might imagine here that you've built a machine to shake the
box, and that you can record the magnitude and direction of
every individual shake it applies, then simply program it to
apply the precise reverse of that sequence.)

If you find this confusing, don't worry. So did the best of physicists throughout the last century, and even today there remain
some puzzling facets to the apparently simple behavior of
highly populous systems under random motion. Let us begin to
unravel some of the puzzles.

A set of one thousand green peas and fifty yellow peas in a
box has an almost uncountable number of configurations or
states available to it. A configuration here means simply a list
of the positions and speeds of all the peas.

The only significant constraint on the possible configurations
or states of the peas in the box is that the total amount of
energy represented by their motion stays roughly constant.
We'll suppose that the box is being shaken with more or less
constant vigor, so that as peas rattle around exchanging energy
with each other, the box as a whole neither loses nor gains
energy in the long term (so that the peas neither come to rest
nor get increasingly agitated, but rattle around indefinitely at
the same sort of rate).

Of all the possible configurations of the system, most will
correspond to what we would call mixed-up states. Imagine
going through all the peas one by one, giving each one a random flick of your fingers to set it moving, and choosing, also at
random, fifty of them to be yellow while the others remain
green. It's pretty unlikely that if you go through this randomizing you will happen to end up with all the yellow peas in one
corner, or with all the peas moving in the same direction, and

so on. We won't begin to try figuring out just how many possible configurations there are for all the peas, let alone what exact fraction represent mixed-up versus unmixed states, but we can be sure there are lots of possible states and that only a few of them represent "special" configurations, such as having the yellow peas all together.

Now we're set. As the peas in the box continue to rattle around, colliding with each other, slowing down, speeding up, changing places, what the whole system is doing is shuffling around pretty much at random from one possible configuration to another. And since the vast majority of these configurations represent mixed-up rather than unmixed boxes of peas, it's no surprise that the system as a whole spends almost all of its time in a mixed-up state (or rather, switching rapidly from one mixed-up state to another). Every now and then, to be sure, it might find itself in a less mixed-up state, but the more orderly the state we're thinking of, the less likely the system is to find itself there. This is the essence of Poincaré's theorem, which says that in the long haul every individual state has the same chance of being visited, but obviously mixed-up states are going to turn up much more often than the less mixed-up or perfectly orderly states, because there's just many more of them (and for any kind of a system of reasonable size, the outnumbering of mixed-up to unmixed states is such that it could take many times the age of the universe for the unmixed states to show up).

In the same vein, it's also no surprise now that an initially orderly state gets rapidly transformed into a mixed-up state, while the reverse rarely happens. Again because of the outnumbering of one kind of state by the other, the system is overwhelmingly more likely to go from order to disorder than the other way around. Random motion among all possible states means that yellow peas easily get mixed into the green peas, but very rarely "mixed out" again.

On the other hand, unmixing is not totally impossible. The reason peas get mixed up is a matter fundamentally of probability, not of the laws of physics; it's not that unmixing is impossible, just that it's highly unlikely. In this sense, the tendency of systems to get mixed up is not an example of absolute irreversibility, because the opposite can in principle happen simply by chance. So if you were thinking, as many classical physicists originally did, that there must be some law of physics that made irreversible processes such as mixing inevitable, you'll be disappointed. It's now understood that irreversibility in the general sense, and even within the strict confines of classical physics, is a probabilistic phenomenon only. Classical physics, it now appears, is not quite as deterministic as it used to be: some things happen by chance and not by absolute decree.

And finally, what about trying to "unmix" the yellow peas by "unshaking" the box? To be precise, you would have to do two things. First, you would need to note down all the positions and speeds of the peas, then go through them individually and exactly reverse their directions of motion; this would set them going in a time-reversed sense through all the collisions they originally went through to reach the point where you decided to stop the clock. Second, you would need to have recorded all the shakes and rattles of the box, so you could apply them too in precise reverse order. In principle, if you could do all this, you could indeed reverse precisely the motion of every single pea in the box and thus bring them gracefully back to their starting point. And if you had worked all this out in advance, you could bring your friends into the room, show them an apparently well-mixed-up box of peas, switch on your programmed box-shaking machine, and have it produce, after a suitable lapse of time and to the amazement of your audience, an unmixed box of peas, with all the yellow ones in one corner, just as they were originally.

Your friends might go so far as to think you had invented a time machine, so implausible is it that shaking up a box of peas would unmix rather than mix them. That might be more exciting than gin rummy!

But you may be thinking that unmixing is not as easy as it has been made to sound. You'd be right. The difficulty is not one of principle but of practice. Suppose you have a configuration of peas and a program of shaking that should unmix the peas and bring them back to the starting point. What are the consequences of being slightly in error with the configuration or the program? It's not hard to see: anything that isn't precisely part of the unmixing scheme represents a tendency toward randomness. Configurations of peas not exactly right are (once more, simply as a matter of probability) more likely to lead you toward randomness—a mixed state—rather than the exactly unmixed state you are aiming for. Similarly, any sequence of rattles and shakes that isn't just right will inevitably lead toward randomness, or mixing.

Probability always works against you. Because there are many more mixed than unmixed arrangements, all errors are overwhelmingly likely to work against you. And once you get off track a little bit (meaning that you are heading in a direction that will not quite bring you back to the unmixed state you'd like to achieve) then in fact your precisely adjusted series of rattles and shakes will no longer be the rattles and shakes that will take you where you want to go. For any given configuration of peas, there is in principle a series of motions that will take you back to an orderly, unmixed state. But that same series of motions applied to a different configuration—even a slightly different one—will be nothing special, just one more random sequence of rattles and shakes of the box. And so it will mix you up instead of unmixing.

In simple terms, when you shake up a box of peas, mixing is not strictly speaking inevitable, but is nevertheless overwhelmingly more likely than unmixing; and if you try by careful design to unmix the peas, your task is not strictly speaking impossible, but is overwhelmingly difficult. And these things hold true despite the fact that all the pea collisions, which are the only physical process actually going on here, are individually and accurately time reversible, and by themselves neither convey nor connote any direction of time.

A brief digression about time

Unmixing the mixed-up peas may be, in practice, difficult or even impossible, but in principle you can think of a series of shakes and rattles applied to the box that would restore them to their original unmixed state. And if you performed this trick in your living room, you might convince your friends that you had invented a time machine. And you might even think, in a sense, that you had, because what you would have achieved is a precise reversal of the motions that had mixed up the peas in the first place.

Is this really some sort of time machine, or just an illusion?

To reverse the mixing of the peas, the setup and the shaking program have to be unimaginably precise. Any small deviation in either the configuration of the peas or the sequence of shakes and rattles will take you off the track, and send you down the path to just another mixed-up box of peas. But in specifying the required precision even to this degree, we have still not taken into account all the potentially confusing factors. Suppose, when you tried to perform the unmixing for your friends, the temperature of the room was a little different from when you had carefully recorded the mixing that you now intend to undo? Might a difference in temperature slightly affect the dynamics of all the collisions, enough to send you down the wrong path and ruin your demonstration?

No physical system is ever truly isolated from the rest of the world, except in our imaginations. Gravity affects the peas in the box; if you shifted the furniture around between mixing up the peas and attempting to unmix, would the resulting tiny changes in the local gravitational field in your living room be enough to throw the unmixing off track?

For real dried peas, and only a thousand of them, the task of unmixing is intricate and delicate, but it's not entirely beyond the bounds of possibility to think about doing such a thing. But if instead we apply the same sort of reasoning to a still more intractable case of irreversibility—let us say, using a wooden spoon to stir a drop of food coloring into a bowl of frosting, in which case we have to think of individual atoms bouncing around, rather than peas—then we begin to think that reversing the seemingly irreversible may not even in principle be possible.

The problem is that as a system gets more complicated (by substituting trillions upon trillions upon trillions of atoms for the mere thousand or so peas we've been dealing with so far) the precision needed to unstir, and the number by which mixed states of frosting plus food coloring outnumber unmixed states, become astronomically large—to the point where we might think that even a change of gravity due to the different relative position of the Sun and Moon is enough to disturb the interaction between atoms so that we are thrown off the "unmixing" track and sent on instead to a merely somewhat different but just as mixed-up state.

If every tiny external influence on the mixing of food coloring and frosting might have a perceptible effect on the attempt to unstir it and bring the food coloring back to a single drop lying on the surface, then it would not be enough simply to have the correct arrangement of atoms and precisely prescribed motion of the wooden spoon. You would have to make sure in addition that every other influence—the temperature of the room, the position of the furniture and the planets—would be the same. And if someone had wandered into the kitchen while you were stirring the bowl, you would have to arrange for that same person to "unwander" back out at the appropriate moment in your unstirring.

In short, to unstir the food coloring out of the frosting you really would have to reverse time, arranging for everything that happened as you stirred to happen in reverse as you unstir. This means that spectators who followed your stirring would, as they watched the unstirring, have to themselves replay their former actions in reverse, and if (perhaps getting a little carried away now) that reversal included not just their overall bodily actions but also the internal motions and dispositions of their constituent atoms, then the reversal would include also a reversal of their brain metabolism, so that they would unobserve the things they had observed originally and unremember whatever they remembered—and by the time this demonstration was done, the spectators would have systematically lost the memory of whatever it was they had been hoping to see happen in reverse. And, therefore, the reversal of the passage of time would be so complete that the participants could in fact not themselves experience it, because they would be part of the self-same reversal.

Let us not pursue this extravagant thought. It's in fact not at all clear quite what one means by a reversal of time in such a case. Simply to unmix the peas or unstir the frosting would be a remarkable accomplishment, but an observer knowledgeable in the behavior of these kinds of complicated systems would not be obliged to think that time had run backwards; instead, it could be an ingenious but explicable demonstration piece of forward-running dynamics.

The interesting point is that physicists use a fundamental but somewhat abstract definition of time in order to think about individual collisions between peas or atoms, and with this kind of argument can show that "perceived" time—which we measure by watching irreversible events such as peas being mixed up or frosting being stirred—tends to follow the same direction as this underlying abstract time. But the connection between the two is not absolute, because "irreversible" things can in principle reverse themselves, by chance or design. The question that's unanswered in all this is what kind of time we, as biological perceiving creatures, primarily sense or understand; is our sense of time tied to the almost perfectly irreversible events in the world around us, or do our minds somehow have access to something more fundamental? That must be a question left to the reader, since no one else knows the answer and it most likely has nothing to do with quantum mechanics.

The defining difference

An electron's spin can be up or down.
A ring of superconducting material can be in one
magnetic state or another.
A box of dried peas can be mixed or unmixed.
A cat can be dead or alive.

The measurement problem in quantum mechanics arises, in a nutshell, because all these statements sound more or less the same. If an electron can be in a simultaneously up-and-down-but-not-really-either spin state—a superposition, as we have come to call it—and if a large object like a superconducting ring can be in a superposition of two distinct magnetic states, why cannot a cat likewise be in a simultaneously dead-and-alive state? And if a cat cannot, in fact, be in such a state, as appears to be the case from everyday experience, what makes it assume, definitely, one state or the other? How does this "collapse" into distinct states, which is the essence of a quantum mechanical measurement, actually come about? Where does the superposition go?

Now that we have considered at length the physics of peas shaken about in a box, we can see that these statements are in fact not all the same. Spin, for an electron, is a defining characteristic. If you know what spin state the particle is in, you know all there is to know as far as that measurement is concerned. Of course, an electron has other measurable properties, such as its momentum and direction of motion, but if we are specifically conducting a quantum measurement of spin, then the two possible results of such a measurement embody complete information about the spin state. Once a measurement is made and a result obtained, there is nothing else to be said. And in the same vein, if we say that before the measurement is made, the electron exists in a superposition of spin states corresponding to the two possible outcomes of the measurement, then we have a full description of the electron.

And likewise, although the superconducting ring contains many paired-up electrons, its magnetic properties derive from the collective behavior of the electrons, which move through the ring in a coherent way. It is the coherence among all the electron pairs that gives rise to the superconductivity (the electrons cooperate rather than compete among themselves to carry the electrical current), and it also means that the overall behavior of the ring can be very simply stated: if you know what magnetic state the ring is in, then you know all there is to know about the underlying behavior of the electrons, because all the electron pairs are essentially doing the same thing; they're all described by a single wavefunction.

But among objects made of lots of electrons or atoms, the superconducting ring is clearly the exception, not the rule. To say that a box full of green and yellow peas is either mixed-up or unmixed is to give a gross characterization of its overall appearance, but not one that tells you very much about the individual disposition of the peas within it. As we saw, there are many possible unmixed states, and many more possible mixed states, and each of those states can encompass wholly different and unrelated individual configurations of peas.

Extending this same sort of analysis to the atoms of a cat, we now see that saying whether a cat is dead or alive does not amount to specifying a "state" of the cat in the same sense that an up or down spin measurement specifies the state of an electron.

There are many possible arrangements of atoms that represent live cats, and many possible arrangements that represent dead cats. Saying whether a particular cat is dead or alive tells us next to nothing about the configuration of all its constituent atoms. Or to put it the other way around, if a physicist with access to divine knowledge were able to specify completely for you the entire quantum state of all the atoms and electrons in a cat, you would not easily be able to deduce from that information whether the cat was actually alive or dead: a list of the positions and velocities and spins of all the particles in a cat doesn't help you very much in knowing what the real, macroscopic cat is up to.

Still, somewhere in all that information the behavior of the cat must reside, since a complete specification of the quantum state of every atom in it includes all possible information you could ever hope to have about the cat. But a cat, unlike a superconducting ring, is a complicated construction. Getting from a detailed knowledge of all its component parts to an understanding of the whole is not, in any practical sense, a feasible matter. For the superconducting ring, coherent behavior of the constituent electrons makes this sort of deduction possible; there is a simple and direct equivalence between the state of the electrons and the magnetic state of the ring itself. But for a cat we cannot make this leap of deduction, and most things in the world are more like cats than like superconducting rings.

Where does this get us? We can legitimately think about the "quantum state" of a cat, if by that we mean a particular and specific arrangement of the cat's entire complement of atoms and electrons, with all relevant quantum properties—spin, momentum, and so on—enumerated. And we can then imagine that some fraction of those states correspond to live cats, and some to dead cats. And, just like the peas rattling around the box, random jiggling around of the atoms in a cat (atoms are always jiggling around in any object that's not at the absolute zero of temperature) ensures that the quantum state of a live cat is endlessly switching from one possible quantum state to another, within the domain of quantum states corresponding to live cats, and similarly for a dead cat.

In short, to treat the "aliveness" or "deadness" of the cat as some kind of macroscopic quantum state, in the same way that we can legitimately think of the magnetic quantum state of a superconducting ring, now becomes a highly dubious proposition. We have glibly talked of live cats and dead cats using the same language and concepts that we became familiar with in talking of up and down electrons. By getting a grip on the complexity of the quantum states of a cat, compared to the simplicity of a quantum state for an electron or even a superconducting ring, we can begin at last to get a grip on the measurement problem itself.

At last, the quantum cat

The quantum state of a cat, we have now realized, means a specification of the exact disposition of every single atom belonging to the cat, along with a specification of the state of every electron associated with every atom. To say merely that a cat is dead or alive is not to specify a quantum state at all. But we can say, on the other hand, that all the possible quantum states of the cat can be divided into one set representing live cats and another set representing dead cats.

We also know that the cat's quantum state, specified with this exactitude, is not a constant thing at all; random motion and interaction of the atoms with each other, not to mention that the cat—a live cat, at least, breathing and eating—is interacting with its environment, make the quantum state of a cat an endlessly changing thing, flickering from one possible quantum state to another. And a dead cat too, even one of those mummified cats found from time to time in the Pyramids, is an inconstant creature: its atoms are jiggling around, and certain unpleasant chemical reactions may be taking place.

On the other hand, a cat cannot pass from any quantum state to any other with complete abandon. We know that a cat will not spontaneously turn into a dog (even though it may be possible to construct, in principle, a plausible lapdog out of the very same atoms that constitute a cat), and we know that dead cats do not spontaneously turn into live ones. (All live cats eventually change into dead ones, of course, but let us focus our attention on healthy cats, given food and water, so that the only cause of transformation from life to death will be some external agent—such as the poison gas released when triggered by the detector that registers the outcome of the Stern-Gerlach measurement of the electron that's in the box with the cat, at Schrödinger's request.)

An obstacle to our reasoning arises here, in the inescapable fact that we do not know how to deduce, from a complete specification of the quantum state of every atom and every electron in a cat, what the actual, real-life cat itself is doing. Even if we knew, in other words, the complete specification of the quantum state of a cat at some particular time, there's no way we could tell whether that state corresponded to a live or dead cat. We may be confident that every quantum state corresponds to a cat that's either dead or alive, but we have no easy way of telling which is which. And in fact it's not clear, simply from the point of view of technical feasibility, whether we could ever devise a mathematical formula that could be applied to all the information contained in the description of a quantum state to say whether the cat is dead or alive.

But, as with the peas in the box, technical complications such as these should not deter us from making some elementary observations. Live cats stay alive, and dead cats stay dead; although the complete quantum description of the cat changes constantly, with incalculable speed and complexity, the deadness or aliveness of the cat is quite evidently a stable, constant property, seemingly immune to these incessant internal readjustments. We can therefore take it as an empirical fact that, although we do not know how to define deadness or aliveness as collective properties of the quantum states of the cat, the distinction is real. All the "live" quantum states of the cat may transform frenetically among themselves, but those states nevertheless have some property in common that we can call "aliveness," which remains constant and stable despite the internal chaos. And the same for dead states.

Now, at last, we are in a position to think clearly about a cat as a quantum measuring device. As per Schrödinger's instructions, we install our cat in a box along with a suitable quantum device; let's use an automatic machine that sends an electron through a Stern-Gerlach magnet, so that if the spin of the electron is "up," the cat stays alive, and if the spin is "down," then alas, it's curtains for kitty. Regardless of all the details, we can say that an up state of the electron is coupled, via all the machinery of the box, to a continuing live state of the cat, while a down state for the electron is coupled to a dead state for the cat.

And the paradox was, in the past, that a "half-up, half-down" electron, which is what obtains when we make a spin measurement on an electron of which we have no prior knowledge, leads simply and directly to a "half-alive, half-dead" state for the cat, which seems meaningless and makes us think no measurement has actually taken place.

But now we know better. Live and dead cat states are not fundamental, constant things at all. Imagine that at the precise moment the electron's spin is measured and the vial of poison is either broken open or left intact, the cat is indeed sent into some strange quantum state consisting of a superposition of dead and alive states, deriving from the superposition of up and down states of the electron. Never mind that we found ourselves unable to say precisely what such a state might look like.

We can, notionally, think of one particular live state—that is, one of the almost innumerable quantum states corresponding to a live cat—and along with it one particular dead state. With those two specific states in mind, we can define (in terms of a mathematically stated wavefunction, at least) a half-alive, half-dead superposition of cat states. To repeat, the state we are trying to imagine here is a superposition of one specific "live cat" state, out of all the possible quantum states corresponding to a live cat, with one specific "dead cat" state out of all the possible quantum states corresponding to a dead cat. In the immediate aftermath of the electron-spin measurement, the cat is indeed placed into a strange superposition of states, both dead and alive but yet neither one thing nor the other.

But then what happens? The cat can't possibly stay in that initial state, because the individual quantum states are not stable. They mix around, one live state transforming constantly into another live state, then another, and the same thing with the dead state. As time goes by, our description of the cat has to change. The single live cat state that we started with, as half of our cat superposition, rattles around among all possible live cat states, so that after a very short time we must describe that half of the cat's wavefunction as a highly changeable mixture of many possible live cat states. That half of the wavefunction becomes spread out, in fact, in the space of all possible live cat states.

And by the same logic, the other half of the cat's wavefunction, the one that started as a single dead state, becomes spread out among all possible dead states.

The cat's wavefunction has evolved, therefore, from a superposition of one live state with one dead state, to a much more complicated superposition of countless live states and countless dead states. So what? This sounds worse, not better, if our aim is to banish superpositions.

But now comes the interesting bit. Originally, at the moment of measurement, the presence of a superposition in the cat's overall state was represented by the presence of a combination of one live state and one dead state; now, after the mere passage of time, that same superposition is represented by a combination of countless live states, each present with some probability, and countless dead states, each also present with some probability. And to make any statement about the cat as a whole, you now have to do a kind of averaging over all those possible, coexisting dead and live states; each individual state is present with some tiny probability, but the actual, macroscopic, real cat behaves as an aggregate of all those possible states.

And the important result is that when you average over all those possible live and dead states, all the pieces representing what would be a genuine superposition of live and dead actual cats cancel out. Microscopically, the individual states are all there; macroscopically, all the mixed live-and-dead bits of the wavefunction manage to negate each other.

This sounds rather mysterious and fortuitous, does it not? But we can see in a picturesque way how it works. Once the measurement has been made, and the cat's wavefunction starts traveling around among all possible individual quantum states, the live and the dead pieces evolve quite independently of each other. That is, the live part of the wavefunction spreads out among all possible live states in a way that's completely unrelated to the way the dead part spreads out among all possible dead states. The two processes are, to use the physicists' term, incoherent.

To find out, at some later time, the probability that the cat is alive, you have to average (loosely speaking) over all the possible live states. The spreading out of the wavefunction among the live states is complicated, but nevertheless not random, because all those live states share, after all, the common property of aliveness. When you average over all these states, you get some number for the probability that the cat is actually alive. Similarly, averaging over the dead states gives you a number for the probability that the cat is dead.

But if you want to figure out the probability that the cat is macroscopically both dead and alive (whatever that means), you have to do a mixed average over all the live and dead states. And because the live states and the dead states are moving around incoherently—that is, independently of each other—the mixed average consists of a lot of bits from all the individual wavefunctions, which end up canceling each other out because they take on all possible values, completely at random.

Schrödinger's cat—the actual, real, macroscopic cat that we see—therefore has some probability of being alive, some probability of being dead, and no probability at all of being both alive and dead at the same time. This vanishing of the probability for the superposed state is known as "decoherence." Any initial coherence between the individual live and dead states that existed at the moment of measurement rapidly disappears because of the incoherent changeability of the live and dead parts of the wavefunction.

What, in practical terms, does this loss of coherence mean? To say that the cat, at the very moment of the electron-spin measurement, was left in a superposed half-alive, half-dead state amounts to saying that the entire cat was set up in some peculiar and unusual coherent state. But this coherent state cannot last. The atoms and electrons jiggle around in their independent ways, and that initial coherence is lost. Because live cat states and dead cat states go their independent ways, what began as a very special state of superposed live-cat-plus-dead-cat evolves almost instantaneously into something which is either dead or alive, but not both at once. The superposition goes away, and we are left with a cat that's either one thing or the other. It is, in fact, what we would think of if we imagined that there is a cat in the box, that the cat is certainly either dead or alive, but that we don't know which until we open the box and find out. That strange quantum superposition (both dead and alive but neither one thing nor the other) has been transformed into simple classical ignorance (either dead or alive).

The key element here is the loss of coherence. In a superconducting ring, physical conditions conspire to ensure that the collective behavior of all the electrons in the ring is coherent, so that an initial state that can be described as a superposed state of two different magnetic states will remain a superposition until something happens to disrupt the collective behavior of the electrons—an external measurement, for example, or simply warming the superconductor above its critical temperature so that superconductivity is lost and the electrons behave independently (incoherently, if you like) as they do in a normal conductor.

In the cat, as in all everyday macroscopic objects, the special kind of coherence that obtains in a superconductor is never there in the first place. The atoms always behave with a good deal of independence, and live quantum states and dead quantum states always evolve in an independent or incoherent manner. And that is why a cat that is set up at one moment in a superposed half-alive, half-dead quantum state will stay that way for only the fleetingest fraction of a second, and will evolve almost instantly into a cat that is genuinely either alive or dead, in the familiar way we understand that phrase.

Schrödinger's cat, in short, does not exist. Or rather, it has an immeasurably short lifetime, once the electron-spin measurement is made, before it evolves spontaneously into a regular, everyday, classical, Newtonian cat.

The ghost of Schrödinger's cat

Can it really be this simple? With the recognition that "live" and "dead" quantum states of a cat are not static but dynamic, that the precise internal specification constantly shuffles from one individual state to another, the measurement problem seems to have gone away. As with the appearance of irreversibility in classical systems, what happens here is essentially a consequence of the large numbers of states involved. Irreversibility in a classical system—the mixing together of our different colored peas, or more powerfully the mixing of a drop of food coloring into a bowl of frosting—involves so rapid and thorough a shuffling around between so many possible internal configurations that all memory of the initial unmixed state is effectively lost. Similarly, even when the quantum cat is set up in a genuine superposition of alive and dead states, the delicacy of adjustment required to sustain that state is blown apart by the incessant shuffling of the cat's state among almost countless possibilities.

But that raises a question similar to one that cropped up in the strictly classical case. If the live and dead quantum states are rattling around so fast among themselves, is it not possible that an initial superposition of half-alive, half-dead states might spontaneously reappear, no matter how rarely and transiently? Or that a perfectly ordinary cat might spontaneously find itself transformed into a half-alive, half-dead cat? Or even that a dead cat might spontaneously transform into a half-alive, half-dead cat, and thence into a genuinely alive cat?

The answer to all these questions is a resounding "yes, but not really."

There is a chance, tiny but calculable, that the box of peas we so vigorously shook up could spontaneously separate itself so that a crowd of yellow peas shows up in one corner, distinct from the green peas elsewhere. There is a chance, much tinier but still calculable, that the molecules of food coloring could accidentally assemble themselves together to re-form the colored droplet standing on the surface of the frosting. There is room even in classical physics for things to happen that seem to violate cause-and-effect. It's just that in any practical sense these things are so unlikely that we can forget about them.

The same is true, and for essentially the same reasons, of the quantum mechanical process of decoherence, which makes initially coherent superposed states rapidly disappear. There's a finite and calculable possibility that decoherence can occasionally act against type, and cause a coherent state to reemerge. To take an extreme example, it's not totally impossible that a dead cat might, as a fantastically unlikely manifestation of decoherence, pop up as a live cat. It's also not impossible that an appropriate mixture of atoms placed in a bucket and stirred might assemble themselves into a living, breathing cat. It's just not the kind of thing we really need to worry about.

But there's a more subtle way that a memory of the initial superposition survives, and one that helps to explain why decoherence, though the idea has been around for a long time, has only in the last decade or so become widely accepted as an essential ingredient in understanding the apparent definiteness and irreversibility of quantum measurements.

Quantum states, whether simple or complicated, evolve according to the equation obtained by Erwin Schrödinger in 1926, and bearing his name. His equation has the property that mathematicians call linearity; what this means is that any two separate solutions of the equation, added together, also form a solution. This has another implication: any solution of the equation that starts out as a sum of two separate solutions will in general remain the sum of two separate pieces, each evolving as they would in isolation. And this has significance for the evolution of superposed states, which can be written mathematically as the sum of separate parts corresponding to the states that are superposed. A sum remains a sum; a superposition remains a superposition.

This seems to contradict what we just learned about decoherence making superpositions go away, and indeed this very point was seized on by John Bell as an argument against the idea that decoherence could be a sound or complete solution to the measurement problem.

The resolution of this apparent contradiction involves another subtlety that arises in going from detailed quantum states (the disposition of all the atoms in a cat) to the observed macroscopic states (the aliveness or deadness of the cat). Of course, we don't know how to express aliveness or deadness in terms of the fundamental quantum states of the cat, but in principle it must be possible; there must be some complicated mathematical construction which, for any given specific quantum state of all the components, will say either yes, the cat is alive, or no, the cat is dead. (If you like, think of a complicated mathematical expression into which you feed all the quantum information about the cat's atoms and electrons, and out of which you get a simple number; if the number comes out positive, the cat's alive; if it's negative, the cat is dead.)

We began the explanation of decoherence with the idea that the cat was in some strange superposed state of deadness and aliveness. But this statement involves two factors: the actual quantum state of the cat, and also the quality or characteristic we are measuring. A superposed state always has this aspect: it has to be expressed in terms of a quantum system along with something about that system that can be measured. For example, think of an electron that has just passed through a Stern-Gerlach magnet and whose spin you know to be "up." That electron is now in a definite "up" state with respect to another vertical Stern-Gerlach magnet, but with respect to a horizontal magnet, the same electron is in a superposed "half-left, half-right" state, because it can come out of either the left or the right channel with equal probability. The electron, therefore, can be in a definite state with respect to one measurement, but an indefinite, superposed state with respect to another.

Back to the cat, which we left just as it had been set up in a superposed state with respect to a dead-or-alive measurement. Decoherence, so it is claimed, causes this state to become rapidly randomized among all possible individual dead cat and live cat quantum states, so that the special coherence of that initial state is lost. But at the same time the linearity of Schrödinger's equation guarantees that the state still consists of equal parts of two different things—a superposition, in other words.

The explanation of this seeming dilemma is that although the initial superposition of states cannot simply be done away with, it must undergo incessant transformation, as the quantum state of the cat rattles around among all the possible internal cat states. What began as a superposition with respect to the dead-or-alive measurement is incessantly and unpredictably transformed into superpositions with respect to other measurements—measurements that we either do not or cannot make.

What might these other measurements be? The deadness or aliveness of the cat, we know, can be represented in principle by some complex mathematical expression based on the detailed quantum disposition of its atoms and electrons. But just as there are trillions upon trillions of possible individual quantum states of the cat, so there are trillions upon trillions of different ways that those dispositions can be combined into macroscopic cat qualities that might, in principle, be measured. One such mathematical expression might tell us, for example, not that the cat is either dead or alive, but that it is either blork or blurp, and a cat that was momentarily in a superposed half-alive, half-dead state might—a few moments later and just for the tiniest fraction of a second—be in a superposed half-blork, half-blurp state. And then, in an instant, it would turn into something else.

The problem is, of course, that we don't know what blork and blurp are, or how to measure them, because they are just some arbitrary combination of the individual quantum states of the cat's atoms and electrons. And, what's more, even if we did have a device that measured blorkness or blurpness, we wouldn't have enough time to make a precise measurement because the superposed state with respect to that quality only lasts the tiniest of moments before decoherence carries it off into something else.

And so, in any practical sense, we never get to see where the superposition went. It flits around, ghostlike, among all the ever-changing internal quantum states of the cat, and like a ghost can never be trapped or incontrovertibly measured.

In which Einstein's Moon is restored

If Schrödinger's cat represents the archetypal quantum measurement, then does our solution to the cat paradox represent an answer to the paradox of quantum measurement in general? That is, can it be proved that all macroscopic objects, such as are used for making measurements, undergo the decoherence process by which mysterious quantum superpositions—neither one thing nor the other but some of both—turn rapidly and (to all intents and purposes) irrevocably into old-fashioned classical mixed states—either this or that, and certainly not both?

The bad news, to be quite candid, is that no one has yet proved in a watertight, rigorous way that decoherence is guaranteed to work for all realistic detecting devices, whether a cat or simply a pointer moving against a scale. But some things can be proved—among them, the solidity of the Moon, which so troubled Einstein. The position of the Moon is marked by its center of mass, which is in essence just the average of the positions of all the atoms that constitute the Moon. Each of those atoms is itself described by a wavefunction, and if, as Einstein used to argue, you think of each atom's wavefunction as marking not precisely the position of that atom but only the position it might have, were some suitable observation to be performed, then you are left to wonder whether the entire Moon, which in quantum mechanical language is now nothing more than a vast assembly of wavefunctions, is really there when no one is looking. Does it take a measurement of some sort to force the Moon to take on a real rather than a potential location in space?

But even when no one is looking at the Moon, it's neither inert nor in total isolation. Its atoms are jiggling around as if connected to each other by little springs, and the surface of the Moon is bombarded by particles and radiation, chiefly from the Sun. A photon from the Sun doesn't collide with the Moon as a whole, but with one of the atoms on the Moon's surface; the wavefunction of that photon and that atom interact and tangle, and the atom that's struck then acquires a little bit of extra energy, which it proceeds democratically to share with its fellow atoms, by means of all the mutual vibrations and interactions among them.

Physically, photons from the Sun do not push the Moon around a whole lot. Quantum mechanically, however, the effect of all those apparently negligible but numerous collisions is to keep the individual wavefunctions of the atoms of the Moon in a state of constant readjustment. The wavefunctions are mixed up, randomized, shuffled around. An individual "quantum state" of the Moon as a whole is a specific set of atomic wavefunctions, and now we see that the Sun's photons mix those wavefunctions around, and so cause the quantum state of the Moon to be in constant flux. And that, it turns out, is enough to put paid to any idea that the Moon could exist in a macroscopic quantum superposition, in which you would have to say it was partly at one position along its orbit, partly at another, but not really in either. This is decoherence at work again: the constant randomizing among possible individual quantum states of the Moon rapidly erases any possible coherent macroscopic superposition of states for the Moon as a whole. It's not so much that the effects of these photon collisions on the Moon are so large, but that maintenance of a coherent superposed state for so large an object as the Moon requires an impossibly exquisite delicacy of adjustment of all the individual atomic wavefunctions that constitute the Moon's overall state. A superposition of states corresponding to the Moon being in two places at once is not impossible, just as a superposed "half-dead, half-alive" cat is not impossible. But both things are so extraordinarily hard to arrange that in practice they never happen, and even if they did momentarily happen they would disappear again almost incalculably fast.

The Moon really is there, after all, when no one's looking. In a general sense, Einstein's comment was correct: quantum mechanics demands that a measurement be made in order for the Moon really to exist at a particular spot. But the new insight afforded by the decoherence argument is that the rain of solar photons onto the Moon's surface is enough of a physical process to constitute a "measurement"—it's enough to get rid of superposed states, which is what we want a measurement to accomplish. No actual observation is required, and the whole process carries on efficiently and relentlessly without any intervention of human action, let alone human consciousness. The world works in its own way, and doesn't need us to look at it.

Even so, it has not been and perhaps cannot be proved that decoherence will always work in the desired way. The task of enumerating all the possible quantum states of the Moon or of a cat is far too daunting to be even conceivably a resolvable task. One must make do with generalities and arguments that establish plausibility rather than proof.

This is not as bad as it might sound. Physicists have long been comfortable with making generalizations about the behavior of atoms in a gas, for example, that allow them to understand the gross nature of a volume of gas—its temperature, pressure, and so on—without having to know what every atom is up to at every possible moment, or even to think that such knowledge can ever be practically obtainable. In the same way, it would be cheering to think that general arguments about the overall dynamics and properties of quantum states can be used to establish that decoherence is an inevitable occurrence, apart from obviously exceptional cases such as superconductors, where special conditions maintain coherence between all the individual quantum states and allow macroscopic quantum superpositions to exist.

But as we said before, the majority of things in the world are more like cats than superconductors. Decoherence is essentially a process of randomization, and in any kind of a large system with complicated inner workings, the sort of randomization needed to make decoherence work tends to occur of its own accord, whereas maintaining coherence takes special care and careful setup. All objects that might be used for making measurements on quantum systems—meters, pointers, flashing lights, cats—must have at least two fundamental properties. First, they must be physically large enough that we can use them as elements of a measuring device; even if the "little end" of a measuring device, the part that's actually in contact with the quantum system to be measured, is some tiny, precise sensor invisible to the human eye, the whole point of measurement is that this sensor has to be connected to the "big end" of the measuring device, which displays a result that we can see. There has to be a pointer, a computer that prints numbers on a piece of paper, an electromagnet that operates a blade that cuts a string that drops a lead brick on our foot—something, in other words, that brings the result of the measurement to our attention.

And second, whatever it is in the measuring device that indicates the result to us must have the potential to exist in a certain number of stable and distinguishable states. The pointer must be able to point in different directions; the computer must be able to print different numbers on the paper; the brick either falls on someone's foot, or it doesn't. Measuring devices of all kinds, in short, must have macroscopic (that is, visible) states that are stable and distinguishable, so that we know some kind of irreversible and perceptible measurement has taken place.

And then the argument for decoherence would be that any such measuring device—composed as it must be of atoms that are jiggling around and interacting and generally being buffeted by each other and by local heat and gravity and noises and people shouting in the corridor outside the laboratory—must be in a constant tizzy of internal motion, endlessly shuffling from one internal quantum state to another, but maintaining nevertheless an appearance of constancy to the outside world. The pointer stays where it is, despite all these random internal motions, unless some specific event happens that causes the pointer to move. A measuring device is useful to us only if it responds reliably and predictably to an input of our choosing.

These observations, it is hoped, are enough to make a case that decoherence will always work. The essential feature of any measuring device—that it has stable, macroscopic states in spite of the constant shuffling among internal quantum states—is precisely the property that allows decoherence to dispose efficiently of any possible macroscopic superposition of states while leaving the result of the measurement intact.

What have we learned?

Let's be optimistic. Decoherence is a general and inevitable property of all complicated systems, and guarantees that quantum superpositions of macroscopic states, as typified by "half-alive, half-dead" cats, never in practice occur. Measurements therefore get made; measuring devices yield definite results, either one thing or another, and never get hung up in indefinite, neither-one-thing-nor-the-other superpositions that mirror the indefiniteness of the thing they are trying to measure.

And because this is an inevitable and universal phenomenon, we no longer have to worry about some supposed influence external to quantum mechanics—whether it's consciousness or the branching of the world into parallel universes—that causes measurements to take place. The enormous appeal of the idea of decoherence is that it follows from quantum mechanics itself, more specifically by applying an appropriately detailed quantum mechanical description to the inner workings of measuring devices. Decoherence may be called a purely technical achievement. It took a long time for physicists to master the understanding of complicated interacting systems with enough rigor and power to be able to make a case—even now, not a proof—that decoherence can solve the measurement paradox that languished so long within the otherwise simple and appealing Copenhagen interpretation of quantum mechanics. But if it's a technical achievement it surely illustrates the truth of Hegel's dictum that quantitative change eventually becomes qualitative; more detailed and accurate understanding of the inner workings of quantum systems has led to a profound change in the way we understand those systems as a whole, and has resolved what seemed to be a genuine logical dilemma.

Does this mean all our worries about interpreting quantum mechanics have been erased? Not exactly . . .

For one thing, classical physics—the time-tested physics of everyday objects, whose validity was not suddenly nullified by the advent of quantum mechanics—depends on more than just the ability to make individual measurements. Classical physics also embodies a traditional idea of objectivity, meaning that whole sets of measurements on a physical system could legitimately and accurately be taken to refer to a unique and reliable underlying "reality." But we've seen over and over, in everything from the simplest Stern-Gerlach measurement of electron spin to the two-slit photon interference experiment to the most ingenious tests of Bell's theorem, that quantum mechanics specifically forbids you from trying to construct a unique reality that underpins a set of measurements. Reality is what you measure it to be, and no more: that's the lesson of the Copenhagen interpretation, and though decoherence may have clarified the meaning of the word "measure" in that statement, it doesn't seem to have gotten us anywhere in understanding the difference between classical and quantum reality.

And we should take note that classical physics still seems to work pretty well, in most areas of physics, and that along with it, a classical view of reality does not cause any trouble except for carefully arranged tests of individual quantum systems. So if classical physics is to be built on a foundation of quantum mechanics, as we would like, then it becomes necessary to explain how apparently solid classical reality arises from the fuzzy unreliability of quantum reality.

To begin with, there's an important result, known as Ehrenfest's theorem, that goes back to the early days of quantum mechanics. What it demonstrates is that an individual wavepacket—the quantum mechanical wavefunction representing a particle such as an electron or a photon—travels around for the most part as if it were a little classical object. The quantum mechanical laws that prescribe the behavior of the wavepacket ensure that the position and momentum that an observer would attribute to the particle represented by the wavefunction behave as if they were governed by the classical laws of dynamics. In other words, even when we think of the position and momentum of the particle not as fundamental classical quantities but as properties derived from the yet more fundamental rules of quantum mechanics, they still behave the same way.

Ehrenfest's theorem holds true almost all the time. It's exactly correct for a wavefunction sailing undisturbed through empty space, and it's almost exactly correct for a wavefunction that's moving through a region where conditions change slowly. It breaks down when the wavefunction is forced, so to speak, to corner on a dime. A wavefunction has a certain physical size to it, and Ehrenfest's theorem fails when some physical disturbance (a magnetic field, for example, or the influence of another wavefunction) has a distinctly different effect on different parts of the wavefunction. Then you can no longer treat the whole wavefunction as a single entity, and the simple correspondence with classical physics breaks down.

But that's exactly what you'd expect. The rule is that a wavefunction can be regarded more or less as a classical object, as long as you are looking at it on a scale where its internal structure can't be discerned. When you look at it on the fine scale, and cannot ignore its structure and physical size, then it behaves like a quantum object, and doesn't necessarily do what you'd expect.

This is a reassuring but limited result. When we think about the Moon in its orbit, we're not thinking about a single object described by a single quantum mechanical wavefunction, but a large and complicated thing made up of lots of individual wavefunctions, which are both doing their own thing and also interacting with each other. What we see, in the form of the classically real, solid, Einsteinian Moon, is some grossly simplified, averaged-out, collective aspect of the behavior of all those internal wavefunctions and their machinations. Is there some generalization of Ehrenfest's theorem that would allow us to say that the quantum mechanical laws governing the behavior of all those individual wavefunctions should reproduce, in sum, what we see as the simple and classical behavior of the actual Moon?

This turns out to be a difficult question, which again was only satisfactorily resolved as physicists learned the techniques that allowed them to understand the dynamical behavior of complicated systems. To cut a long story short, the answer to the question is yes, with two provisos.

First, there's some unavoidable fuzziness in defining the gross classical properties of an object such as the Moon—its position and momentum, its rotation, mass, shape, and so on—in terms of the individual wavefunctions of all the atoms in the Moon. In effect, for any classical property of the Moon, there are many possible quantum definitions, all more or less but not quite equivalent. The position of the Moon (defined by the position of its center of mass or the sharpness of its surface) turns out to have a little bit of quantum uncertainty to it.

Second, there's a small quantum probability that the Moon might do something totally unexpected and inexplicable, by classical standards. It might, for example, suddenly dematerialize and reappear on the other side of its orbit. Or turn inside out. These things are not impossible, but you would have to wait around for times exponentially longer than the age of the universe for them to happen. They're unlikely in the same sense, and for the same sort of reason, that quantum superpositions of macroscopic objects are unlikely: they require all the individual wavefunctions of the object to momentarily and spontaneously conspire in some classically forbidden collective action.

It may sound a little alarming that weird behavior of this sort is not completely ruled out, but in fact we've gotten used to this kind of near-impossibility even in classical physics. As with our example of food coloring being mixed into a bowl of frosting, there's always some tiny but not quite zero probability that continued random stirring could spontaneously "unmix" the frosting. It's also possible for boiling water in a kettle to spontaneously cool down, through some random conjunction of atomic motions, and give its heat up to the surrounding atmosphere.

Quantum mechanics reproduces the classical physics of large objects with a little bit of slack, therefore, but it's a kind of slack, arising from the statistics of large numbers, that we're already accustomed to. Classical determinism is slightly imperfect, and so it remains when it's derived as a consequence of the underlying quantum mechanical behavior of all the constituent parts.

Now, where are we? Decoherence can explain how measurements occur, so that in the world of large objects we need deal only with definite states and can forget about fuzzy quantum superpositions. On top of that, if we have a definite classical state, defined as some complicated collective property of numerous constituent wavefunctions, then we can be sure that the laws of quantum mechanics governing those wavefunctions reproduce, on the large scale, the classically expected behavior of the overall object, except for some tiny but to all intents and purposes negligible chance of some weird quantum thing happening. It seems, therefore, as if classical physics has been derived from quantum mechanics, and that in the process properties looking very much like classical determinism and objectivity have arisen. Quite a surprise, you might think, considering how unclassical, indeterministic, and unobjective quantum mechanics seems to be.

But things are not quite as simple as that.

What haven't we learned?

The decoherence "solution" to the measurement problem has been dismissed by some as not a solution at all. Decoherence says that measuring devices generate definite results, not quantum superpositions. But it doesn't tell you which of the various possible results of any individual measurement you are going to get. Probability still reigns. Is this a serious omission in what was looking like a happy ending to our story?

Not really. What decoherence does is to show how the strange and literally unreal quantum superpositions are never encountered in practice. Decoherence makes them go away, leaving a mixture of definite and classically recognizable states uncontaminated by the "half-dead, half-alive" states that have no correspondence with anything ever observed in our large-scale world.

What decoherence doesn't do is tell you whether any particular cat will be dead or alive, any particular electron's spin "up" or "down." But since decoherence is a phenomenon rooted in standard quantum mechanics and nothing else, it can't possibly tell you the outcome of individual measurements. If we persist with our belief that quantum mechanics is the fundamental and complete basic theory of nature, we have to accept that no such prediction can ever emerge. Quantum mechanics is a probabilistic theory. It says that if you measure the spin of an electron of unknown provenance, the result will be up half the time and down the other half of the time. No further information can be extracted from quantum mechanics. Decoherence explains why in fact you get either an up or a down result, and why the "half-up, half-down" result that was seemingly predicted by earlier and simpler analyses never comes to pass in any tangible way. But if quantum mechanics cannot tell you which particular result, up or down, you are going to get in any one experiment, there's no way decoherence, which is "merely" a technical elaboration of the quantum mechanics of complicated systems, can produce an answer.

Decoherence therefore completes the Copenhagen interpretation of quantum mechanics, by making the process of measurement a real physical phenomenon rather than a decree from Bohr. Niels Bohr knew that measurements in fact took place in physics laboratories all around the world, but he couldn't explain how, and so simply declared that they happened. This was always a genuine flaw with his interpretation, but decoherence resolves it. Measurements happen, and you don't need anything besides quantum mechanics to explain how. Nevertheless, the measurement itself remains an act of probability, and all the strange aspects of quantum mechanics—nonlocality, the inability to define an underlying reality that all can agree on—remain to be wrestled with. If you still think that, in the end, there ought to be some way of knowing which result of any particular measurement is going to occur, then you are going to have to look beyond quantum mechanics for that knowledge. But if you accept that quantum mechanics is all there is, then decoherence provides as much of an answer to the measurement problem as you can hope for.

Nevertheless, with the decoherence argument in place, we can see how the fundamentally indeterminate nature of quantum mechanics leads to an almost exactly deterministic large-scale world, in which measurements, once made, are fixed forever, and in which effect follows cause according to the tested and reliable prescriptions of classical physics. To insist now on looking for some sort of deterministic theory that "explains" quantum mechanics, in the way that Bohm and other hidden-variable fans had wished, is to search for a foundation that's no longer needed. If quantum mechanics, indeterminate as it is, builds up into a nigh-on perfectly deterministic world, why add a new layer of determinism in the form of a theory supporting quantum mechanics?

Any such theory would give the logical structure of the world an oddly three-layered form: at bottom would be some sort of hidden-variable determinism—a determinism which we can never directly apprehend but which underlies the apparently probabilistic nature of quantum mechanics. On this second layer, quantum mechanics, we then use decoherence to build a third layer—a macroscopic world which is not exactly deterministic, but which for all intents and purposes can be treated that way. But in such a theoretical structure, the macroscopic determinism we see and rely on (the third layer) has nothing at all to do with the furtive determinism in the first layer that was supposed to "explain" quantum mechanics. Hidden determinism would create the apparently probabilistic form of quantum mechanics, and that probabilistic but still logical theory can account, all by itself, for the classical determinism that controls what Einstein called "the world of our senses."

You can never prove that Bohm's theory is wrong, because at heart it is a mathematical manipulation of standard quantum theory. But if the main aim of Bohm's revision of quantum mechanics was to restore determinism to the world, the issue has become moot. Quantum mechanics can give you as much determinism as anyone actually needs, so the insistence of Bohm, and for that matter Einstein, that there must be something "underneath" quantum mechanics has become empty. Some people may still wish for such a theory, but they can no longer point to any solid practical reason for doing so.

The last (or first) mystery

Decoherence inevitably happens in a large system built of quantum components: its individual quantum states rattle around at random, disposing of all the strange quantum superpositions that depend on almost impossibly precise coherence between all the constituent quantum states. Making those superpositions go away is what makes measurements happen, and it all happens without human intervention. It's a property of large systems in general, not of some specific "act of measurement" that has to be distinguished in some mysterious way from other straightforward physical processes. There's no need of human intervention, still less of human consciousness. It makes "measurement," in the quantum mechanical sense, an independent and understandable physical phenomenon.

What's more, decoherence sifts out from the random buzzing of quantum states a few overall properties that we use to recognize and define physics. We rely, in describing the world, on just a tiny number of broadly defined properties—tiny, at least, in comparison to the huge number of quantum states roiling beneath our ability to see and detect them. Size, position, speed; color, texture, hardness; solidity and fluidity; smell, taste, sound. You can think of more, but compared to the trillions of potential qualities related to the underlying quantum structure of these objects, the handful of things that we notice is small indeed. Decoherence lets these properties exist stably despite the transience and fuzziness of internal quantum states, and the fact that they are selected out in this way gives them classical meaning, making them reliable and deterministic in a way that individual quantum states are not.

It would be agreeable to imagine, therefore, that the universe as a whole owes its apparent solidity and objectivity to the effects of decoherence. Supposing that the universe began as a "big bang" that must ultimately be described as a quantum event, we can think of all the components of the universe flying apart, changing, evolving, according to the underlying rules of quantum mechanics, but we can conjecture also that the chaotic and incessantly busy interaction between all the components amounts to a continuous process of cosmic decoherence that erases quantum fuzziness and presents to us a universe of well-defined (because "decohered") states and properties. Even though it may be fundamentally a quantum phenomenon, the universe thus acquires a classical reliability and evolves according to what look like classical rules of cause-and-effect.

It would be nice to think so. Can it be true?

So far, we've begun to see how decoherence allows individual measurements to produce definite results. But there's more to classical physics than that. The essence of the classical view of the world is not simply that measurements give results, but that all possible measurements that can be carried out on some object or system produce a whole set of mutually consistent results referring, as Einstein emphasized, to a single unambiguous reality.

In quantum physics, we have to take each measurement as it comes, and we have to beware of thinking that different and incompatible measurements made on the same system must necessarily yield consistent results. That's the whole point of the EPR argument, in all its forms, and of Bell's theorem. There is no single "underlying" quantum reality to which all measurements can be referred.

But in classical physics, life is much simpler. There's no such thing as the idea of "incompatible" measurements in the classical world. Faced with any physical system, we can make any measurements we like, in the full expectation that we can later put all the results of our measurements together and compile a distinct and unambiguous picture of the thing we were measuring. There's no uncertainty principle in classical physics; we can measure whatever we want. But if the classical world has quantum foundations, where does the uncertainty principle go?

Part of the answer is that the same properties of decoherence that make classically observable qualities stand out from the busy randomness of the underlying quantum states also make those qualities compatible with each other. What distinguishes macroscopic classical properties is that they are stable and reliable despite the ever-changing nature of the quantum states from which they are built, and that immunity to the constant interchange among quantum states turns out to mean that classical properties are not subject to any kind of uncertainty principle. To put it the other way around, if you tried to define two would-be classical properties of a macroscopic system that were linked by some sort of uncertainty principle (so that measuring one would inhibit you from measuring the other, and vice versa), then you would find that these two properties would not, in fact, be stabilized by decoherence, and would not therefore be true classical properties at all.

That's a helpful argument, but it's still somewhat back to front. Given a recognized classical property, we can reasonably hope to show from quantum mechanical first principles that it behaves in the way we expect classical properties to behave. But ideally, if we want to take with full weight the idea that classical physics flows inevitably from quantum mechanics, we would like to turn that argument around. We would like to prove that the intricate quantum mechanics of large systems necessarily and inevitably gives rise to the classical properties we know and love.

This remains an unfulfilled hope. After the universe was born, and before humans appeared on the scene, we would like to think that all kinds of quantum processes went on that ultimately caused atoms and stars and galaxies and planets to arise from the primordial chaos. And even though we might imagine that, in different notional universes, stars and galaxies and planets might end up in different physical arrangements, we would like to think that overall there was a certain inevitability to the proceedings. We would like to think, in short, that the decoherence of the initial quantum state of the universe created solidity in a generally predictable way.

However, we are by no means yet in a position to say that decoherence necessarily gives rise to a universe that has just the classical properties and behavior that we recognize. Is it possible that in some other universe, a different set of properties—classical, in the technical sense, but unfamiliar to our eyes—arose through the action of decoherence? At every step, as we say, decoherence erases quantum superpositions but does not and cannot choose between different possible outcomes of a quantum measurement. Is it inevitable that all possible outcomes, in the end, still lead to a universe that looks like our own or like a modest variation upon it? Or is it possible that some entirely different history might have been played out, if some apparently insignificant interaction at some early moment had given rise to a different possible outcome?

And when creatures such as ourselves appear in the universe, other questions arise. We can choose to measure different things. We can measure the polarization state of a photon with respect to this angle or that; we can measure the spin of an electron in an up-down or left-right sense, or anything in between; we can measure the position or the momentum of a particle, or some limited combination of the two. And once we have made such a measurement, we set in motion a chain of events that becomes irrevocable. Depending on the outcome of an experiment, a memorable paper might get published in a scientific journal, or a cat may die. The paper can't later be unpublished; the cat can't be restored to life.

Any quantum measurement, or series of measurements, can

set in motion a chain of classical events, one thing following another in familiar manner. But once one chain of events happens, other possible chains of events cannot. Decoherence guarantees that a chain of events rather than a continuously ill-defined stream of quantum possibilities actually takes place. But it doesn't tell us which chain of events is going to happen. Probability has not been erased; measurements can have several different outcomes, and we cannot predict which.

Because decoherence makes measurement a physical process rather than a mysterious and external act, as it was in the original Copenhagen interpretation, we no longer have to think that human intervention is somehow necessary to make a measurement "real." On the other hand, human intervention still plays a role to the extent that we have the freedom to decide which measurement is made in the first place.

As long as we are thinking about measuring a quantum system upon which we, the independent observer, can bring to bear any piece of measuring equipment that we can devise, there is no problem with any of this. But when we think of the universe as a whole, with a single quantum origin, it would seem that there are no "independent observers." If every act of measurement sets into motion a possible chain of subsequent events, and thereby renders other possibilities nugatory, then we must face the fact that we ourselves, through the long history of cosmic and planetary and biological evolution, have arrived at the present moment through the prior action of a whole string of earlier "measurements" that themselves made some possibilities happen but put an end to others. And if we suppose that even our states of mind are at some level quantum states (our brains work because electric signals zip from neuron to neuron, and electric signals are ultimately quantum phenomena), then do we conclude that our "choice" to measure a vertical rather than a horizontal spin is the consequence of a mental state, describable in quantum terms, that arose as a product of prior quantum states decohering their way to present reality since the beginning of the universe? And must we, in the end, treat our own states of mind as states that "decohere," so that the processes that we call "observing a particular outcome" or "making a decision" should be regarded as the almost instantaneous shuffling of previously superposed mental states into distinct rather than mixed "beliefs"?

This is an old problem in new guise. In the heyday of classical physics it seemed that every physical event in the universe must be determined absolutely and predictably from previous events, because the rule of cause-and-effect was inviolable. But then everything in the universe must have been preordained from the outset, so that the absolute determinism of classical physics apparently negated the possibility of free will. No one ever resolved that conundrum.

Quantum mechanics took away absolute determinism, because there's an element of probability in every interaction between particles. Decoherence, in principle, makes this loss of determinism more manageable by, we presume, sifting out from the full and detailed quantum evolution of the universe a smaller number of possible histories corresponding to sequences of events obeying cause-and-effect once more, because we've now understood that cause-and-effect in the familiar sense survives as a consequence of the collective behavior of large quantum systems.

But these are possible histories, not one specific history. And our evolution as perceiving, thinking creatures is part of that history. No one has resolved this conundrum—or, for that matter, genuinely clarified what we need to resolve. If quantum mechanics, applied to the universe in its entirety and with decoherence working its effects, allows many possible cosmological histories, is there some additional rule that picks out one particular history? Would all these possible histories seem, to our perceptions, equally possible (in the sense of conforming to our expectations of how large-scale classical systems should behave)? Or does the fact that we and our brains have evolved along with the universe mean that what we perceive as classical properties are attuned, by that common evolution, to be precisely those properties that we are capable of sensing, so that in some other cosmological history we would perceive as a chaotic and indefinite mess what the inhabitants of that universe found orderly and explicable?

If the contradiction between determinism and free will for classical physics was never resolved, it doesn't follow that a loss of determinism, in the form of quantum probabilities for physical events, makes the contradiction go away. It remains a question to which no one knows the answer. Whether it's even a question whose answer lies within the realm of physics cannot be said. Quantum physics, classical physics, theology—some questions seem eternal.

Will we ever understand quantum mechanics?

But we do, don't we? As an intellectual apparatus that allows us to figure out what will happen in all conceivable kinds of situations, quantum mechanics works just fine, and tells us whatever answers we need to know. Finally we understand how measurements get made, and with that addition the Copenhagen philosophy espoused by Niels Bohr and adopted, with greater or lesser degrees of understanding, by the majority of working physicists is complete. We need only a modicum of intellectual self-discipline, so that we refrain from asking questions that cannot be simultaneously answered, and all's well.

But, of course, understanding in general parlance entails as much of a visceral feeling as an intellectual one—perhaps more. We understand something when it makes sense to us; things make sense because they fit into some sort of idea, no matter how loosely conceived or defined, of the how the world works. Quantum mechanics clearly does not fit into any picture that we can obtain from everyday experience of how the world works. It estranges us from our sense of what the world is like, throws us off balance. Many physicists have been unable to resist the feeling that quantum mechanics is wrong, or at least incomplete, for precisely this reason. Here, for example, are the words of Edwin T. Jaynes, a great figure in statistical physics and information theory:

> It is pretty clear that present quantum theory not only does not use—it does not even dare to mention—the notion of a "real physical situation." Defenders of the theory say that this notion is philosophically naive, a throwback to outmoded ways of thinking, and that recognition of this constitutes deep new wisdom about the nature of human knowledge. I say that it constitutes a violent irrationality, and that somewhere in this theory the distinction between reality and our knowledge of reality has become lost, and the result has more the character of medieval necromancy than of science.

So there's your choice: traditional classical objectivity, or else medieval necromancy! If not perfect order, then complete anarchy!

Physics, and the rest of science, grew up with a belief in objective reality, that the universe is really out there and that we are measuring it. As long as physics could retain this philosophy, there was no reason to question it. And the longer the belief was retained, the more it came to seem as if it must be an essential part of the foundation of physics, perhaps the most essential part of all.

Then quantum mechanics came along and destroyed that notion of reality. Experiment backs up the axioms of quantum mechanics. Nothing is real until you measure it, and if you try to infer from disparate sets of measurements what reality really is, you run into contradictions. It simply can't be done. To many physicists this seemed like heresy against science itself, but as the decades have gone by a new realization emerges: science still works, quantum mechanics and all. The foundations have not crumbled, the walls have not crashed to the ground.

A true believer might take heart from this and conclude that objective reality must still be there somewhere, beneath quantum mechanics. That's what Einstein believed, and what motivated Bohm to put together his hidden variables theory. But an equally valid conclusion, strictly in terms of logic alone, is that if quantum mechanics is the foundation of physics, and if quantum mechanics does not embody an objective view of reality, then evidently an objective view of reality is not essential to the conduct of physics. This is Bohr's position, in essence; his Copenhagen philosophy is a systematic way of dealing with quantum mechanics but at the same time avoiding any unwarranted or unnecessary assumptions about what's "real."

The new development is that quantum mechanics, despite its lack of an objective reality, nevertheless gives rise to a macroscopic world that acts, most of the time, as if it were objectively real. Aspect's experiments, showing that nature does not conform to Bell's theorem, amounted to a demonstration that the nonobjectivity of quantum mechanics could be brought into the open in a carefully designed and executed experiment. But this takes special effort, and such demonstrations do not occur by chance in the world around us.

And so, almost paradoxically, we can believe in an objective reality most of the time, because quantum mechanics predicts that the world should behave that way. But it's because the world behaves that way that we have acquired such a profound belief in objective reality—and that's what makes quantum mechanics so hard to understand.

Can you ride a bicycle? Ah, but do you understand how to ride a bicycle? If you don't know how, the task looks absurd and impossible, and once you know how, it's impossible to explain to someone else, in words and pictures, how to do it. You just keep at it and eventually you get it. Then you can ride a bicycle, and it ceases to baffle you.

In terms of understanding, riding a bicycle is quite the opposite from quantum mechanics. Viscerally, we develop a level of comfort on a bike, but intellectually it remains a puzzling business: physicists have analyzed the mechanics of bicycles, and physiologists have worried over our simple human ability to maintain balance, but there is no simple scientific account of how we stay aboard. We have a visceral understanding but not an intellectual one. With quantum mechanics, even for most physicists, it's the other way around: we can train our intellects and learn how to use quantum mechanics successfully, but it never quite feels comfortable.

Could we ever understand quantum mechanics in the visceral as well as the intellectual sense? The problem is, we never get the chance to experience it directly, and thereby obtain a direct, personal experience of its many strangenesses. Even the physicists who think about its theoretical problems and perform the subtle experiments we have talked about are dealing with the subject at a remove, in terms of mathematized concepts or the statistics of detected electrons. These are not directly the creatures of the quantum world. In fact, the true inhabitants of the quantum world—wavefunctions—are by definition inaccessible to us. We see an outcome only when a wavefunction interacts with a measuring device, and any attempt to think of the wavefunction itself as a real physical thing—like a wave on the ocean or a packet of electromagnetic energy—soon leads into trouble.

So now, if you ask whether we will ever understand quantum mechanics, the answer should perhaps be no, never, because it is a world we cannot experience, and experience is the only thing that can give understanding.

But this is too bleak. Quantum mechanics does make sense, after all, and the Copenhagen interpretation works. Looking for ways to disguise quantum mechanics, to dress it up in classical styles, is never going to work: you always have to trade one puzzle for another, getting rid of indeterminacy only to find instead multiple universes or undetectable guide waves or some such contrivance.

Instead, one must learn to take quantum mechanics at face value and accept that its strange qualities are manageable. It will never conform to our empirical view of what the world should look like, but it constitutes a legitimate world of its own, and we must respect that. Perhaps that's the most profound lesson of all: in quantum mechanics nature is, at the most fundamental level, genuinely unknowable, but despite that, the world at large, the world of which quantum mechanics is the foundation, can be known and understood. Einstein, seeking higher authority for his objections to quantum mechanics, was always saying that "God does not play dice" and that "the Good Lord is subtle, but He is not malicious." And finally an exasperated Bohr admonished Einstein, "Stop telling God what to do!"

Bibliography and notes

To bring myself up to date with the niceties of quantum mechanics I relied heavily on two recent graduate-level textbooks:

Quantum Theory: Concepts and Methods by Asher Peres (Boston: Kluwer Academic Publishers, 1993); and
The Interpretation of Quantum Mechanics by Roland Omnés (Princeton, N.J.: Princeton University Press, 1994).

The first is a technical textbook, which emphasizes, however, in a way that many textbooks do not, the logical problems associated with linking formal quantum mechanical arguments to actual experimental predictions. The second is mathematically somewhat less demanding, but deals rigorously with the problems of logic and meaning that arise in making sense of quantum mechanics. Omnés himself originated many of the arguments set out in his book.

I have indicated in the following notes the ideas I have taken or adapted from these two books, and others. Any errors and misinterpretations are mine, of course.

John Bell's pioneering work in quantum mechanics is collected in *Speakable and Unspeakable in Quantum Mechanics* (Cambridge: Cambridge University Press, 1987). Although this is a volume of research papers, some are aimed at a general audience, and Bell's clarity of expression (unrivaled among commentators on the subject) allows nonphysicist readers to derive insight from his writings.

Although I am unpersuaded by David Bohm's "hidden variables" philosophy, his ideas are influential and have attracted many others to his cause. *The Undivided Universe* by Bohm and Basil Hiley (New York: Routledge, 1993) is the most recent presentation of the hidden variables approach. The book also contains hints of a wider philosophy of the world that Bohm built on his interpretation of quantum mechanics, which I have not delved into here, because it goes well beyond the subject of quantum mechanics itself, and because in any case I find it obscure to the point of incomprehensibility.

A collection of essays in honor of David Bohm, *Quantum Implications* (New York: Routledge, 1987), contains some further elaborations and extrapolations of Bohm's ideas, as well as some perspectives from commentators with different viewpoints.

Many authors have presented accounts of the elements of quantum mechanics, at various levels of sophistication. Two standard accounts of quantum mechanics for the general reader are *Quantum Reality* by Nick Herbert (New York: Anchor Press, 1985) and *In Search of Schrödinger's Cat: Quantum Physics and Reality* by John Gribbin (New York: Bantam, 1984). Both are now somewhat out of date, but give the essentials in an accurate way. Herbert digs into both scientific and philosophical issues, and gives equitable accounts of a number of different views, but (perhaps unfairly) I cannot quite take seriously an author who refers to Immanuel Kant as a "reality researcher." Gribbin is openly a member of the anti-Copenhagen school, and opts instead for the "many worlds" interpretation. In his recent book, *Schrödinger's Kittens and the Search for Reality* (Boston: Little, Brown, 1995), Gribbin describes newer experimental results that exemplify the weirdness of quantum mechanics. He is still anti-Copenhagen, but now plumps for the so-called "transactional interpretation" (see below, "The possibility of simultaneity").

A couple of useful semitechnical books (that is, aimed at nonscientists but including a few diagrams and equations) are Alastair Rae's *Quantum Physics: Illusion or Reality?* (Cambridge: Canto, 1994; originally published by Cambridge University Press, 1986) and J. C. Polkinghorne's *The Quantum World* (Princeton, N.J.: Princeton University Press, 1984).

In the following notes, I provide references for some specific ideas as well as a few technical comments and asides that would have impeded the main text. Most of the material in the first half of the book is quite standard, in terms of technical content, and can be found in the books mentioned above, or indeed in any undergraduate quantum mechanics text.

The mystery of the other glove: Bell (*Speakable and Unspeakable*, chap. 16) has the example of Professor Bertlmann, who always wears socks of a different color. Gloves struck me as an informative alternative, because right-handed and left-handed gloves are distinctly different—so that one cannot imagine any sort of "in-between" glove—and yet undeniably form a pair.

In which things are exactly . . . : The significance of the Stern-Gerlach experiment is detailed by Peres, p. 14 et seq., who explains more fully why classical physics cannot explain the results Stern and Gerlach obtained.

Block that metaphor!: The notion of electron spin arose (in 1925, due to G. Uhlenbeck and S. Goudsmit) in attempts to account for fine details of the spectrum of light emitted and absorbed by atoms. See Abraham Pais, *Inward Bound* (New York: Oxford University Press, 1986), p. 254. In a magnetic field, the energy of atomic electron orbital changes slightly, depending which spin orientation the electron adopts. This leads to a splitting of single spectral lines into two, a phenomenon known as the Zeeman effect.

Polkinghorne (*The Quantum World,* p. 60) also uses the idea of a Stern-Gerlach magnet measuring an electron's spin, without delving into the technical complications I have briefly mentioned here.

More elaborately, an electron has half a unit of quantum mechanical spin, and in a magnetic field must adopt one of two possible spin alignments, $+\frac{1}{2}$ or $-\frac{1}{2}$, corresponding, for example, to "up" or "down" in a Stern-Gerlach measurement. Particles with larger spins also exist. A beam of particles with one unit of spin would emerge from a Stern-Gerlach magnet in three paths, corresponding to spin alignments of $+1$, 0, and -1 with respect to the field; these might correspond to measurements labeled "up," "undeflected," and "down." A particle with $\frac{3}{2}$ spin must adopt values $+\frac{3}{2}$, $+\frac{1}{2}$, $-\frac{1}{2}$, and $-\frac{3}{2}$ with respect to a magnetic field, and so on.

Enter the photon: The development of radiation theory, and its significance for the emergence of quantum theory, is discussed at great length by E. T. Whittaker, *A History of the Theories of Aether and Electricity* (New York: Dover, 1989).

So photons are really real . . . : Whittaker, *A History of the Theories of Aether and Electricity,* again provides much more detail and history.

Is it or isn't it?: Two-slit interference experiments with whole atoms were done by various groups in the early 1990s. See, for example, *Science News,* vol. 140, 7 Sept. 1991, p. 158.

Which way did the photon go?: M. O. Scully, B.-G. Englert, and H. Walther have described an ingenious "which way" experiment, and claim they can determine the presence of a particle in one path or the other without affecting in any important way the particle's dynamics (*Nature,* vol. 351, 9 May 1991, p. 111). This is held to be an ideal experiment, in which you can know the particle's route without in any way disturbing it. The result, nevertheless, is that if you detect a particle, you no longer see interference, as quantum mechanics requires. Whether the detection is quite as delicate as the authors claim, and by what physical effect the interference pattern disappears, continue to be debated (*Science News*, vol. 145, 19 Feb. 1994, p. 118; *Nature*, vol. 377, 19 Oct. 1994, p. 584). But the debate here is not whether quantum mechanics works, but precisely how it works.

No, but really . . . : "Delayed choice" experiments have been made much of by John Wheeler and others, although the idea has earlier beginnings; see Omnés (pp. 456–63), who admits to some puzzlement over what issue such experiments are meant to illuminate. Some accounts of delayed choice experiments have unfortunately adopted the attitude that the physicist's choice of what to measure, after the experiment has begun, has the effect of "changing the past." See, for example, *Newsweek*, 19 June 1995, p. 69.

How to make money . . . : For more about Soros see, for example, two articles in *The New Republic*, 10 and 17 Jan. 1994.

The importance of being rigorous: Wheeler's observation is given, for example, by Herbert, *Quantum Reality*, p. 18.

Psychophysics . . . : The ill-formulated idea that consciousness has something to do with making measurements happen has had a surprisingly long life in the minds of otherwise hard-headed physicists, perhaps because von Neumann seemed to believe it, and because also it puts the measurement problem in a place where we are unable to investigate it further. But it seems to have been left by the wayside. Peres (p. 26) mentions the idea in a sentence, only to dismiss it, while Polkinghorne (*The Quantum World*, pp. 65–67) and Rae (*Quantum Physics*, chap. 5) dismiss it at somewhat greater length.

Does the Moon really exist?: For Einstein's question about the Moon see *Subtle Is the Lord: The Science and Life of Albert Einstein*, by Abraham Pais (New York: Oxford University Press, 1982), p. 5.

The fatal blow?: The original, and still widely quoted, paper by Einstein, Podolsky, and Rosen was published in *Physical Review*, vol. 47 (1935), p. 777. The much-used phrase "elements of physical reality" originated here, although it represents a concept with a long but more or less unexamined history in physics up to this point.

A new spin . . . : The flavor of Bohm's intellectual journey toward Copenhagen and then away again can be discerned in his contribution (chap. 2) to *Quantum Implications*. His book *Quantum Theory* (Englewood Cliffs, N.J.: Prentice Hall, 1951) contains the adaptation to spin measurements of the earlier EPR proposal.

In which Einstein is caught . . . : This version of an EPR experiment, and the way that it puts old Einstein and young Einstein at odds, is taken from Omnés, chapter 9.

Whose reality . . . : The idea that observers somehow "create" reality, and that all observers are engaged in a sort of mutual but apparently unconscious conspiracy to keep the universe going, is hinted at in the last pages of Herbert's *Quantum Reality*, and is taken to an extreme by, for example, Danah Zohar in *The Quantum Self* (New York: Morrow, 1990) and *The Quantum Society* (New York: Morrow, 1994). The appeal of these ideas seems to arise from enlisting fundamental physics in support of a quasi-religious program that has us all part of a "holistic" cosmic enterprise, as in the old Coca-Cola commercials that wanted to teach the world to sing in perfect harmony.

In which Niels Bohr is obscure . . . : Bohr's words, and Bell's admission of his failure to understand them, are in an appendix to chapter 16 of Bell, *Speakable and Unspeakable*. Some of the difficulties here arise from Bohr's use of the word "influence," which seems to suggest that what the observer chooses to do somehow affects the system that is about to be measured. A more neutral statement would be that the observer is obliged to choose between different possible measurements that might be made, and the choice determines the kind of information that results. One may charitably suppose that Bohr was attempting to persuade his critics by adopting what he thought was their language.

And how many universes . . . : Everett's "many worlds" idea was published in *Reviews of Modern Physics*, vol. 29 (1957), p. 454. The proposal generally merits a mention in treatments of the meaning of quantum mechanics, but little of a technical nature has been written about it, beyond the original discussion. This lends weight to the suspicion that once you've said what the idea is, you've said all there is to say about it.

Omnés (p. 347) particularly emphasizes the criticism that Everett's proposal not only multiplies universes, but does so on numerous, seemingly inconsequential occasions.

Indeterminacy as illusion: Technical aspects of Bohm's theory are derived from David Bohm and Basil Hiley, *The Undivided Universe*.

David Z. Albert (*Scientific American*, May 1994, p. 58) offers a philosophical defense of Bohm's ideas. But having castigated the Copenhagen interpretation because it does not permit you to know all you would like to know, Albert seems quite happy to admit the intrinsically undetectable quantum potential as an essential ingredient of the hidden variables philosophy. As always, it's not a matter of right and wrong, but of what you find least objectionable.

In which seeming virtues . . . : For technical reasons, spin cannot be pictured in Bohm's theory by imagining the electron literally as a tiny spinning top. See chapter 9 of David Bohm and Basil Hiley, *The Undivided Universe*. Instead, we are supposed to think of spin as a "context-dependent circulatory motion of [electrons'] trajectories" (p. 220), a phrase I am unable to decipher.

You can push it around . . . : Feynman's words, which originated in his much-admired *Lectures in Physics*, can also be found in the posthumous compilation *Six Easy Pieces* (Reading, Mass.: Addison-Wesley, 1995), p. 117.

Act II: Putting Reality to the Test: Hardy's comment on James is mentioned, for example, by Anthony West in *H. G. Wells: Aspects of a Life* (London: Penguin, 1985), p. 49.

A new angle on EPR: Bell (*Speakable and Unspeakable*, pp. 145–46) emphasizes the point that by concentrating on an EPR experiment in which the two spin detectors were either aligned or perpendicular, physicists were too easily beguiled by the possibility of a deterministic account of the results.

Fun with algebra: My algebraic explanation of Bell's theorem is lifted from Peres (pp. 164–65), who bases his account on a treatment of the problem by J. Clauser, M. Horne, A. Shimony, and R. Holt (*Physical Review Letters*, vol. 23 [1969], p. 880). Bell's original theorem (reproduced as chapter 2 of Bell, *Speakable and Unspeakable*) arose from the case of three spin detectors, two in one path and one in the other. With two detectors in both electron paths, the algebra becomes a little more direct. Nevertheless, the idea is undoubtedly Bell's, and the Clauser-Horne-Shimony-Holt version entails no conceptual novelties.

A genuinely innovative development, however, came from consideration of spin measurements on three or four particles which are created in a state of known overall spin, and which are therefore subject to EPR-type correlations. With more than two particles, it is possible to obtain some direct algebraic contradictions between different sets of measurements, rather than the contradictions with an average of a series of measurements, as in Bell's theorem and its variants. The multiparticle states are, however, somewhat hard to understand without a more detailed command of quantum mechanics. See Peres, p. 152.

And the answer is . . . : The results of Aspect's remarkable experiments were published in the paper by A. Aspect, J. Dalibard, and G. Roger, *Physical Review Letters*, vol. 49 (1982), p. 1804, and have been confirmed by others, using different experimental techniques. See Peres, p. 166, and Omnés, p. 406. The confirmation of quantum mechanical predictions in these experiments is often given only a brief mention in textbooks because it is, after all, a confirmation. Nevertheless, the extent to which this work brought quantitative clarity to a previously somewhat metaphysical debate can hardly be underestimated.

In which reality . . . : My illustration of the generality of the arguments leading to Bell's theorem was stimulated by chapter 13 of Bell, *Speakable and Unspeakable*; see also chapter 16, p. 152. Bell, in these examples, has a curious preoccupation with health statistics in the French towns of Lille and Lyons (birthrate in one case, incidence of heart attacks in the other) which he uses to illustrate the extent of possible correlations between similar data in separate places. Chapter 13 also includes his enumeration of the reasons Bell's theorem might be disobeyed in nature.

Bell's observation that any hidden variables approach resolves the EPR dilemma "in the way which Einstein would have liked least" is on p. 11 of *Speakable and Unspeakable*.

The possibility of simultaneity: J. G. Cramer's "transactional interpretation" of quantum mechanics was presented in *Reviews of Modern Physics*, vol. 58 (1986), p. 647, and is embraced by John Gribbin at the end of his book *Schrödinger's Kittens*. Technically, because Maxwell's equations for the propagation of light are second-order differential equations in time, they possess symmetrical "forwards" and "backwards" solutions. Schrödinger's equation, however, is only first order in time, so does not of its own accord yield up the symmetrical forward and backward halves of the wavefunction that Cramer needs. Cramer argues that both pieces are physically plausible if one allows that the familiar Schrödinger equation is, in effect, one of a pair of similar equations that descend from a common, fully relativistic parent equation for the wavefunction. But this is not spelled out in any detail.

This may also be the place to mention some experiments which allegedly show photons going from point A to point B in less time than something traveling at the speed of light would need (A. Steinberg, P. Kwait, and R. Chiao, *Physical Review Letters*, vol. 71 [1993], p. 708; also *Scientific American,* Aug. 1993, p. 52). In these experiments, photons "tunnel" quantum mechanically through a barrier that, in classical terms, is impenetrable to light. Picturesquely, one may say that the spread-out wavefunction of a photon on one side of the barrier is able to insinuate itself through the barrier in a way that classical electromagnetic radiation cannot, and that the wavefunction that sneaks through can cause a photon to appear on the other side of the barrier.

In this case, it is not right to say that a photon travels through the barrier; rather, a photon disappears on one side, and another photon reappears on the other. The effect depends crucially on the photon's wavefunction being extended in space, and in fact the answer one gets for the travel time "through" the barrier depends on which part of the emerging wavefunction is presumed to be causally linked to which part of the incoming one. Such niceties make it hard to define uncontroversially exactly what is meant by the overall travel time and speed, and the significance of the apparent faster-than-light result remains debatable. See R. Landauer in *Nature*, vol. 365 (1993), p. 692.

Experimental results such as these are, in any case, highly specific to their circumstances, and have no obvious connection with the nonlocality exhibited in EPR experiments.

An engineer, a physicist, and a philosopher . . . : This is an old anecdote whose provenance I can't remember. You can change the characters to suit your purpose (an experimental physicist, a theoretical physicist, and a mathematician; a Protestant, a Catholic, and a Jesuit).

Penrose's words are from his contribution to *Quantum Implications*, p. 106.

The one true paradox: The genuinely troubling nature of measurement, in the Copenhagen interpretation, is emphasized by Omnés, pp. 81–92, and by Peres, pp. 373–74.

Can a quantum superposition be seen?: T. Clark, in chapter 7 of *Quantum Implications*, describes macroscopic quantum measurements on superconducting systems. See also chapter 10 of Omnés.

Like peas in a box: Peres (chap. 11) discusses irreversibility in the classical dynamics of complex systems, with a view to understanding the connection with complex quantum systems. He also talks about truly chaotic systems, which I have left alone because, in many respects, quantum chaotic systems are somewhat less interesting than classical ones.

A brief digression about time: My vague thoughts here are partly inspired by Peres's comments (p. 346) on the practical difficulty of truly isolating any system from the universe at large.

At last, the quantum cat: Decoherence is touched on by Peres, p. 387 et seq., and dissected at much greater length by Omnés, chapter 7.

The technical details of decoherence are somewhat more intricate than my words here can convey. Decoherence itself is not so hard a process to understand; what is more difficult is that one cannot really understand it without understanding in a more precise way the mathematical properties of wavefunctions, and how measurement probabilities are derived from them. The averaging of the wavefunction that I refer to here is meant to indicate the process by which a measurement probability is obtained from a wavefunction, which includes an averaging over all microscopic quantum states that contribute to the same macroscopic state. The vanishing of the contributions from the superposition terms in this averaging is not altogether straightforward. It comes in large part from the incoherent evolution of the separate terms in the wavefunction, but also, and importantly, from the fact that the individual live and dead quantum states correspond to macroscopically distinct live and dead states for the whole cat. This latter property shows up in the orthogonality, as mathematicians call it, of live and dead wavefunctions, which contributes to their cancellation in the averaging process.

The ghost of Schrödinger's cat: Bell's argument against the utility of decoherence as an explanation of measurement, on the grounds that superposition, once established, must always remain, is reproduced as chapter 6 of Bell, *Speakable and Unspeakable*. Omnés (pp. 304–15) argues in turn against Bell's criticism.

There Omnés gives a different but related example than the one I describe of the practical impossibility of measuring a superposition in a macroscopic system. He defines a particular internal state of the system that, by construction, is guaranteed to remain at all times in a superposition; he then shows that to make an unambiguous measurement of that particular state, you would need a measuring device bigger than the whole universe. The reason is that any macroscopic system has a huge number of barely distinguishable internal quantum states, and to measure any of them incontrovertibly, you need a device that can translate one of those internal states into a macroscopic response (a "measurement," in other words). But then the macroscopic states of the putative measuring device have to outnumber, loosely speaking, the internal quantum states of the thing you are measuring. . . . The salient point is that one can define, formally, a "measurable" property of a quantum system that turns out, in a practical sense, not to be measurable at all.

In which Einstein's Moon is restored: Omnés, pp. 297–99 and 319–22.

What have we learned?: Ehrenfest's theorem is on pp. 43–44 of Omnés, who discusses at great length, in his chapter 6, how the concepts of classical physics can be related to fundamental quantum states. It should be emphasized that this is a distinct question from the measurement problem. The justification of "classicism" amounts to showing that, if you start with a certain complex quantum state, or set of states, corresponding to some classical state (such as a baseball moving with a particular speed and direction), then the laws of quantum mechanics cause those underlying states to evolve so as to produce, at some later time, a classical baseball doing just what Newton would have predicted from his laws for the classical dynamical properties of the ball. Because the precepts and laws of quantum mechanics and classical mechanics are so different, this is by no means a straightforward demonstration.

What haven't we learned?: The suggestion that decoherence doesn't really solve the measurement problem is given, for example, by A. Leggett in chapter 5 of *Quantum Implications*, p. 97 in particular, and by A. Rae in his review of Omnés's book in *Physics World* (Institute of Physics, United Kingdom), Feb. 1995, p. 54. Omnés offers a general response to this sort of criticism in his chapter 12, where he presents his views on what we are to make of the nature of fundamental reality, accepting that the classical view no longer holds and that probability cannot be banished. Here, I confess, I begin to lose track of what Omnés is saying and where he hopes to go.

On p. 401, Omnés emphasizes the needlessness of trying to base quantum mechanics on a hidden deterministic foundation, if visible determinism can be shown to follow from quantum mechanics alone.

The last (or first) mystery: I have largely avoided in this book the "cosmological" aspects of quantum mechanics. As long as one thinks of quantum systems and the devices that measure them as independent pieces of the physical world (though all functioning, of course, according to the same physical laws), one can define and attack the measurement problem fairly directly. But once it's admitted that the universe itself began as a single quantum system, from which all else—measurers and things that get measured—descends, then one stumbles into a swamp of seemingly circular questions and answers, many of which have been around, in one form or another, since long before quantum mechanics came on the scene.

M. Gell-Mann and J. Hartle, in *Complexity, Entropy, and the Physics of Information* (Redwood City, Calif.: Addison-Wesley, 1990), outline their program, unfulfilled so far, for how one might seek to understand the emergence of classical properties in a quantum universe. See also an article by W. Zurek, *Physics Today*, Oct. 1991, p. 36, and comments by Omnés, p. 502.

Will we ever understand quantum mechanics?: Jaynes is quoted by Scully et al. (see "Which way did the photon go?" above), from which I have taken this extract.

Bohr's possibly apocryphal admonition to Einstein is quoted with slightly different wording by Rae, *Quantum Physics*, p. 48.

Index